U.S. Department of Justice

Bureau of Alcohol, Tobacco, Firearms and Explosives

Enforcement Programs and Services

I0488205

ATF

Federal Explosives Law and Regulations

2012

U.S. Department of Justice

Bureau of Alcohol, Tobacco,
Firearms and Explosives

Office of the Director

APR 2 7 2012

Washington, DC 20226

www.atf.gov

Dear Federal Explosives Licensee or Permittee:

We are pleased to provide you with the latest edition of the Federal Explosives Law and Regulations. We hope this guide is useful for you as we continue our shared goal of ensuring the safe and secure storage of explosive materials. ATF values each and every licensee and permittee as a vital partner in pursuit of this mission, and we appreciate your cooperation.

This edition of the Federal Explosives Law and Regulations reflects changes and developments in response to the Safe Explosives Act of 2002. Federal explosives laws have evolved greatly since the enactment of Title XI of the Organized Crime Act of 1970, and we at ATF want to keep industry members informed about the developments that affect them and their businesses.

Ensuring the safety of explosive materials can be accomplished through proper construction of magazines, adherence to the Tables of Distances, and appropriate housekeeping. Safety is also achieved through effective employee training, proper handling procedures, and compliance with all State and local requirements.

Security of all explosive materials helps protect Americans from violent crime and is an essential tool in the war against terrorism. ATF combats the criminal diversion or misuse of explosive materials by verifying required records, ensuring the use of approved locking systems, and conducting background checks on industry applicants and their employees. Licensees and permittees assist us with our mission through internal controls, best practices and ATF-industry security publications.

Your continued vigilance is the most effective means of safety and security. ATF appreciates your cooperation and support in helping us meet our shared goals. As always, if we can be of assistance, please do not hesitate to contact us.

Sincerely yours,

B. Todd Jones
Acting Director

Table of Contents

Regulation of Explosives

Public Law 91-452, Approved October 15, 1970 (as Amended)

[Note: Any reference to the Internal Revenue Code of 1954 refers to the Internal Revenue Code of 1986 (§ 2, Public Law 99–514, 100 Stat. 2085, October 22, 1986.)]

Purpose

Sec. 1101. The Congress hereby declares that the purpose of this title is to protect interstate and foreign commerce against interference and interruption by reducing the hazard to persons and property arising from misuse and unsafe or insecure storage of explosive materials. It is not the purpose of this title to place any undue or unnecessary Federal restrictions or burdens on law-abiding citizens with respect to the acquisition, possession, storage, or use of explosive materials for industrial, mining, agricultural, or other lawful purposes, or to provide for the imposition by Federal regulations of any procedures or requirements other than those reasonably necessary to implement and effectuate the provisions of this title.

Sec. 1102. Title 18, United States Code, is amended by adding after chapter 39 the following chapter:

Chapter 40. Importation, Manufacture, Distribution and Storage of Explosive Materials

Editor's Note: The sections of law set out herein were added by Public Law 91–452, Title XI, § 1102(a), Oct. 15, 1970, 84 Stat. 952-959, and remain un-changed unless otherwise footnoted.

§ 841. Definitions

As used in this chapter

(a) **"Person"** means any individual, corporation, company, association, firm, partnership, society, or joint stock company.

(b) **"Interstate or foreign commerce"** means commerce between any place in a State and any place outside of that State, or within any possession of the United States (not including the Canal Zone) or the District of Columbia, and commerce between places within the same State but through any place outside of that State. "State" includes the District of Columbia, the Commonwealth of Puerto Rico, and the possessions of the United States (not including the Canal Zone).

(c) **"Explosive materials"** means explosives, blasting agents, and detonators.

(d) Except for the purposes of subsections (d), (e), (f), (g), (h), (i), and (j) of section 844 of this title, **"explosives"** means any chemical compound mixture, or device, the primary or common purpose of which is to function by explosion; the term includes, but is not limited to, dynamite and other high explosives, black powder, pellet powder, initiating explosives, detonators, safety fuses, squibs, detonating cord, igniter cord, and igniters. The Attorney General shall publish and revise at least annually in the Federal Register a list of these and any additional explosives which he determines to be within the coverage of this chapter. For the purposes of subsections (d), (e), (f), (g), (h), and (i) of section 844 of this title, the term "explosive" is defined in subsection (j) of such section 844.

(e) **"Blasting agent"** means any material or mixture, consisting of fuel and oxidizer, intended for blasting, not otherwise defined as an explosive: Provided, that the finished product, as mixed for use or shipment, cannot be detonated by means of a numbered 8 test blasting cap when unconfined.

(f) **"Detonator"** means any device containing a detonating charge that is used for initiating detonation in an explosive; the term includes, but is not limited to, electric blasting caps of instantaneous and delay types, blasting caps for use with safety fuses and detonating cord delay connectors.

(g) **"Importer"** means any person engaged in the business of importing or bringing explosive materials into the United States for purposes of sale or distribution.

(h) **"Manufacturer"** means any person engaged in the business of manufacturing explosive materials for purposes of sale or distribution or for his own use.

(i) **"Dealer"** means any person engaged in the business of distributing explosive materials at wholesale or retail.

(j) **"Permittee"** means any user of explosives for a lawful purpose, who has obtained either a user permit or a limited permit under the provisions of this chapter.

(k) **"Attorney General"** means the Attorney General of the United States.

(l) **"Crime punishable by imprisonment for a term exceeding one year"** shall not mean (1) any Federal or State offenses pertaining to antitrust violations, unfair trade practices, restraints of trade, or other similar offenses relating to the regulation of business practices as the Attorney General may by regulation designate, or (2) any State offense (other than one involving a firearm or explosive) classified by the laws of the State as a misdemeanor and punishable by a term of imprisonment of two years or less.

(m) **"Licensee"** means any importer, manufacturer, or dealer licensed under the provisions of this chapter.

(n) **"Distribute"** means sell, issue, give, transfer, or otherwise dispose of.

(o) **"Convention on the Marking of Plastic Explosives"** means the Convention on the Marking of Plastic Explosives for the Purpose of Detection, Done at Montreal on 1 March 1991.

(p) **"Detection agent"** means any one of the substances specified in this subsection when introduced into a plastic explosive or formulated in such explosive as a part of the manufacturing process in such a manner as to achieve homogeneous distribution in the finished explosive, including

(1) Ethylene glycol dinitrate (EGDN), $C_2H_4(NO_3)_2$, molecular weight 152, when the minimum concentration in the finished explosive is 0.2 percent by mass;

(2) 2,3-Dimethyl-2,3-dinitrobutane (DMNB), $C_6H_{12}(NO_2)_2$, molecular weight 176, when the minimum concentration in the finished explosive is 0.1 percent by mass;

(3) Para-Mononitrotoluene (p-MNT), $C_7H_7NO_2$, molecular weight 137, when the minimum concentration in the finished explosive is 0.5 percent by mass;

(4) Ortho-Mononitrotoluene (o-MNT), $C_7H_7NO_2$, molecular weight 137, when the minimum concentration in the finished explosive is 0.5 percent by mass; and

(5) any other substance in the concentration specified by the Attorney General, after consultation with the Secretary of State and the Secretary of Defense, that has been added to the table in part 2 of the Technical Annex to the Convention on the Marking of Plastic Explosives.

(q) **"Plastic explosive"** means an explosive material in flexible or elastic sheet form formulated with one or more high explosives which in their pure form has a vapor pressure less than 10-<4> Pa at a temperature of 25 °C, is formulated with a binder material, and is as a mixture malleable or flexible at normal room temperature.

(r) **"Alien"** means any person who is not a citizen or national of the United States.

(s) **"Responsible person"** means an individual who has the power to direct the management and policies of the applicant pertaining to explosive materials.

(t) **Indian Tribe.** The term "Indian tribe" has the meaning given the term in section 102 of the Federally Recognized Indian Tribe List Act of 1994 (25 U.S.C. 479a).

(Added Pub. L. 91–452, title XI, Sec. 1102(a), Oct. 15, 1970, 84 Stat. 952; amended Pub. L. 104–132, title VI, Sec. 602, Apr. 24, 1996, 110 Stat. 1288; Pub. L. 107–296, title XI, Secs. 1112(e)(1), (3), 1122(a), Nov. 25, 2002, 116 Stat. 2276, 2280; Pub. L. 111–211, title II, Sec. 236(b), July 29, 2010, 124 Stat. 2286.)

§ 842. Unlawful Acts

(a) **It shall be unlawful for any person—**

(1) to engage in the business of importing, manufacturing, or dealing in explosive materials without a license issued under this chapter;

(2) knowingly to withhold information or to make any false or fictitious oral or written statement or to furnish or exhibit any false, fictitious, or misrepresented identification, intended or likely to deceive for the purpose of obtaining explosive materials, or a license, permit, exemption, or relief from disability under the provisions of this chapter;

(3) other than a licensee or permittee knowingly—

(A) to transport, ship, cause to be transported, or receive any explosive materials; or

(B) to distribute explosive materials to any person other than a licensee or permittee; or

(4) who is a holder of a limited permit—

(A) to transport, ship, cause to be transported, or receive in interstate or foreign commerce any explosive materials; or

(B) to receive explosive materials from a licensee or permittee, whose premises are located outside the State of residence of the limited permit holder, or on more than 6 separate occasions, during the period of the permit, to receive explosive materials from 1 or more licensees or permittees whose premises are located within the State of residence of the limited permit holder.

(b) It shall be unlawful for any licensee or permittee to knowingly distribute any explosive materials to any person other than—

   (1) a licensee;

   (2) a holder of a user permit; or

   (3) a holder of a limited permit who is a resident of the State where distribution is made and in which the premises of the transferor are located.

(c) It shall be unlawful for any licensee to distribute explosive materials to any person who the licensee has reason to believe intends to transport such explosive materials into a State where the purchase, possession, or use of explosive materials is prohibited or which does not permit its residents to transport or ship explosive materials into it or to receive explosive materials in it.

(d) It shall be unlawful for any person knowingly to distribute explosive materials to any individual who:

   (1) is under twenty-one years of age;

   (2) has been convicted in any court of a crime punishable by imprisonment for a term exceeding one year;

   (3) is under indictment for a crime punishable by imprisonment for a term exceeding one year;

   (4) is a fugitive from justice;

   (5) is an unlawful user of or addicted to any controlled substance (as defined in section 102 of the Controlled Substances Act (21 U.S.C. 802));

   (6) has been adjudicated a mental defective or who has been committed to a mental institution;

   (7) is an alien, other than an alien who—

      (A) is lawfully admitted for permanent residence (as defined in section 101(a)(20) of the Immigration and Nationality Act);

      (B) is in lawful nonimmigrant status, is a refugee admitted under section 207 of the Immigration and Nationality Act (8 U.S.C. 1157), or is in asylum status under section 208 of the Immigration and Nationality Act (8 U.S.C. 1158), and—

         (i) is a foreign law enforcement officer of a friendly foreign government, as determined by the Attorney General in consultation with the Secretary of State, entering the United States on official law enforcement business, and the shipping, transporting, possession, or receipt of explosive materials is in furtherance of this official law enforcement business; or

         (ii) is a person having the power to direct or cause the direction of the management and policies of a corporation, partnership, or association licensed pursuant to section 843(a), and the shipping, transporting, possession, or receipt of explosive materials is in furtherance of such power;

      (C) is a member of a North Atlantic Treaty Organization (NATO) or other friendly foreign military force, as determined by the Attorney General in consultation with the Secretary of Defense, who is present in the United States under military orders for training or other military purpose authorized by the United States and the shipping, transporting, possession, or receipt of explosive materials is in furtherance of the authorized military purpose; or

      (D) is lawfully present in the United States in cooperation with the Director of Central Intelligence, and the shipment, transportation, receipt, or possession of the explosive materials is in furtherance of such cooperation;

   (8) has been discharged from the armed forces under dishonorable conditions;

   (9) having been a citizen of the United States, has renounced the citizenship of that person.

(e) It shall be unlawful for any licensee knowingly to distribute any explosive materials to any person in any State where the purchase, possession, or use by such person of such explosive materials would be in violation of any State law or any published ordinance applicable at the place of distribution.

(f) It shall be unlawful for any licensee or permittee willfully to manufacture, import, purchase, distribute, or receive explosive materials without making such records as the Attorney General may by regulation require, including, but not limited to, a statement of intended use, the name, date, place of birth, social security number or taxpayer identification number, and place of residence of any natural person to whom explosive materials are distributed. If explosive materials are distributed to a corporation or other business entity, such records shall include the identity and principal and local places of business and the name, date, place of birth, and place of residence of the natural person acting as agent of the corporation or other business entity in arranging the distribution.

(g) It shall be unlawful for any licensee or permittee knowingly to make any false entry in any record which he is required to keep pursuant to this section or regulations promulgated under section 847 of this title.

(h) It shall be unlawful for any person to receive, possess, transport, ship, conceal, store, barter, sell, dispose of, or pledge or accept as security for a loan, any stolen explosive materials which are moving as, which are part of, which constitute, or which have been shipped or transported in, interstate or foreign commerce, either before or after such materials were stolen, knowing or having reasonable cause to believe that the explosive materials were stolen.

(i) It shall be unlawful for any person—

   (1) who is under indictment for, or who has been convicted in any court of, a crime punishable by imprisonment for a term exceeding one year;

   (2) who is a fugitive from justice;

   (3) who is an unlawful user of or addicted to any controlled substance (as defined in section 102 of the Controlled Substances Act (21 U.S.C. 802));

(4) who has been adjudicated as a mental defective or who has been committed to a mental institution;

(5) who is an alien, other than an alien who—

(A) is lawfully admitted for permanent residence (as that term is defined in section 101(a)(20) of the Immigration and Nationality Act);

(B) is in lawful nonimmigrant status, is a refugee admitted under section 207 of the Immigration and Nationality Act (8 U.S.C. 1157), or is in asylum status under section 208 of the Immigration and Nationality Act (8 U.S.C. 1158), and—

(i) is a foreign law enforcement officer of a friendly foreign government, as determined by the Attorney General in consultation with the Secretary of State, entering the United States on official law enforcement business, and the shipping, transporting, possession, or receipt of explosive materials is in furtherance of this official law enforcement business; or

(ii) is a person having the power to direct or cause the direction of the management and policies of a corporation, partnership, or association licensed pursuant to section 843(a), and the shipping, transporting, possession, or receipt of explosive materials is in furtherance of such power;

(C) is a member of a North Atlantic Treaty Organization (NATO) or other friendly foreign military force, as determined by the Attorney General in consultation with the Secretary of Defense, who is present in the United States under military orders for training or other military purpose authorized by the United States and the shipping, transporting, possession, or receipt of explosive materials is in furtherance of the authorized military purpose; or

(D) is lawfully present in the United States in cooperation with the Director of Central Intelligence, and the shipment, transportation, receipt, or possession of the explosive materials is in furtherance of such cooperation;

(6) who has been discharged from the armed forces under dishonorable conditions;

(7) who, having been a citizen of the United States, has renounced the citizenship of that person to ship or transport any explosive in or affecting interstate or foreign commerce or to receive or possess any explosive which has been shipped or transported in or affecting interstate or foreign commerce.

(j) It shall be unlawful for any person to store any explosive material in a manner not in conformity with regulations promulgated by the Attorney General. In promulgating such regulations, the Attorney General shall take into consideration the class, type, and quantity of explosive materials to be stored, as well as the standards of safety and security recognized in the explosives industry.

(k) It shall be unlawful for any person who has knowledge of the theft or loss of any explosive materials from his stock, to fail to report such theft or loss within twenty-four hours of discovery thereof, to the Attorney General and to appropriate local authorities.

(l) It shall be unlawful for any person to manufacture any plastic explosive that does not contain a detection agent.

(m) (1) It shall be unlawful for any person to import or bring into the United States, or export from the United States, any plastic explosive that does not contain a detection agent.

(2) This subsection does not apply to the importation or bringing into the United States, or the exportation from the United States, of any plastic explosive that was imported or brought into, or manufactured in the United States prior to the date of enactment of this subsection [enacted April 24, 1996] by or on behalf of any agency of the United States performing military or police functions (including any military reserve component) or by or on behalf of the National Guard of any State, not later than 15 years after the date of entry into force of the Convention on the Marking of Plastic Explosives, with respect to the United States.

(n) (1) It shall be unlawful for any person to ship, transport, transfer, receive, or possess any plastic explosive that does not contain a detection agent.

(2) This subsection does not apply to—

(A) the shipment, transportation, transfer, receipt, or possession of any plastic explosive that was imported or brought into, or manufactured in the United States prior to the date of enactment of this subsection [enacted April 24, 1996] by any person during the period beginning on that date and ending 3 years after that date of enactment; or

(B) the shipment, transportation, transfer, receipt, or possession of any plastic explosive that was imported or brought into, or manufactured in the United States prior to the date of enactment of this subsection [enacted April 24, 1996] by or on behalf of any agency of the United States performing a military or police function (including any military reserve component) or by or on behalf of the National Guard of any State, not later than 15 years after the date of entry into force of the Convention on the Marking of Plastic Explosives, with respect to the United States.

(o) It shall be unlawful for any person, other than an agency of the United States (including any military reserve component) or the National Guard of any State, possessing any plastic explosive on the date of enactment of this subsection [enacted April 24, 1996], to fail to report to the Attorney General within 120 days after such date of enactment the quantity of such explosives possessed, the manufacturer or importer, any marks of identification on such explosives, and such other information as the Attorney General may prescribe by regulation.

(p) Distribution of Information Relating to Explosives, Destructive Devices, and Weapons of Mass Destruction.

(1) Definitions. In this subsection—

(A) the term "destructive device" has the same meaning as in section 921(a)(4);

(B) the term "explosive" has the same meaning as in section 844(j); and

(C) the term "weapon of mass destruction" has the same meaning as in section 2332a(c)(2).

(2) Prohibition. It shall be unlawful for any person—

(A) to teach or demonstrate the making or use of an explosive, a destructive device, or a weapon of mass destruction, or to distribute by any means information pertaining to, in whole or in part, the manufacture or use of an explosive, destructive device, or weapon of mass destruction, with the intent that the teaching, demonstration, or information be used for, or in furtherance of, an activity that constitutes a Federal crime of violence; or

(B) to teach or demonstrate to any person the making or use of an explosive, a destructive device, or a weapon of mass destruction, or to distribute to any person, by any means, information pertaining to, in whole or in part, the manufacture or use of an explosive, destructive device, or weapon of mass destruction, knowing that such person intends to use the teaching, demonstration, or information for, or in furtherance of, an activity that constitutes a Federal crime of violence.

(Added Pub. L. 91–452, title XI, Sec. 1102(a), Oct. 15, 1970, 84 Stat. 953; amended Pub. L. 100–690, title VI, Sec. 6474(c), (d), Nov. 18, 1988, 102 Stat. 4380; Pub. L. 101–647, title XXXV, Sec. 3521, Nov. 29, 1990, 104 Stat. 4923; Pub. L. 103–322, title XI, Secs. 110508, 110516, Sept. 13, 1994, 108 Stat. 2018, 2020; Pub. L. 104–132, title VI, Sec. 603, title VII, Sec. 707, Apr. 24, 1996, 110 Stat. 1289, 1296; Pub. L. 106–54, Sec. 2(a), Aug. 17, 1999, 113 Stat. 398; Pub. L. 107–296, title XI, Secs. 1112(e)(3), 1122(b), 1123, Nov. 25, 2002, 116 Stat. 2276, 2280, 2283; Pub. L. 108–177, title III, Sec. 372, Dec. 13, 2003, 117 Stat. 2627.)

§ 843. Licenses and User Permits

(a) An application for a user permit or limited permit or a license to import, manufacture, or deal in explosive materials shall be in such form and contain such information as the Attorney General shall by regulation prescribe, including the names of and appropriate identifying information regarding all employees who will be authorized by the applicant to possess explosive materials, as well as fingerprints and a photograph of each responsible person. Each applicant for a license or permit shall pay a fee to be charged as set by the Attorney General, said fee not to exceed $50 for a limited permit and $200 for any other license or permit. Each license or user permit shall be valid for not longer than 3 years from the date of issuance and each limited permit shall be valid for not longer than 1 year from the date of issuance. Each license or permit shall be renewable upon the same conditions and subject to the same restrictions as the original license or permit, and upon payment of a renewal fee not to exceed one-half of the original fee.

(b) Upon the filing of a proper application and payment of the prescribed fee, and subject to the provisions of this chapter and other applicable laws, the Attorney General shall issue to such applicant the appropriate license or permit if—

(1) the applicant (or, if the applicant is a corporation, partnership, or association, each responsible person with respect to the applicant) is not a person described in section 842(i);

(2) the applicant has not willfully violated any of the provisions of this chapter or regulations issued hereunder;

(3) the applicant has in a State premises from which he conducts or intends to conduct business;

(4) (A) the Attorney General verifies by inspection or, if the application is for an original limited permit or the first or second renewal of such a permit, by such other means as the Attorney General determines appropriate, that the applicant has a place of storage for explosive materials which meets such standards of public safety and security against theft as the Attorney General by regulations shall prescribe; and

(B) subparagraph (A) shall not apply to an applicant for the renewal of a limited permit if the Attorney General has verified, by inspection within the preceding 3 years, the matters described in subparagraph (A) with respect to the applicant; and

(5) the applicant has demonstrated and certified in writing that he is familiar with all published State laws and local ordinances relating to explosive materials for the location in which he intends to do business;

(6) none of the employees of the applicant who will be authorized by the applicant to possess explosive materials is any person described in section 842(i); and

(7) in the case of a limited permit, the applicant has certified in writing that the applicant will not receive explosive materials on more than 6 separate occasions during the 12-month period for which the limited permit is valid.

(c) The Attorney General shall approve or deny an application within a period of 90 days for licenses and permits, beginning on the date such application is received by the Attorney General.

(d) The Attorney General may revoke any license or permit issued under this section if in the opinion of the Attorney General the holder thereof has violated any provision of this chapter or any rule or regulation prescribed by the Attorney General under this chapter, or has become ineligible to acquire explosive materials under section 842(d). The Attorney General's action under this subsection may be reviewed only as provided in subsection (e)(2) of this section.

(e) (1) Any person whose application is denied or whose license or permit is revoked shall receive a written notice from the Attorney General stating the specific grounds upon which such denial or revocation is based. Any notice of a revocation of a license or permit shall be given to the holder of such license or permit prior to or concurrently with the effective date of the revocation.

(2) If the Attorney General denies an application for, or revokes a license, or permit, he shall, upon request by the aggrieved party, promptly hold a hearing to review his denial or revocation. In the case of a revocation, the Attorney General may upon a request of the holder stay the effective date of the revocation. A hearing under this section shall be at a location convenient to the aggrieved party. The Attorney General shall give written notice of his decision to the aggrieved party within a reasonable time after the hearing. The aggrieved party may, within sixty days after receipt of the Attorney General's written decision, file a petition with the United States court of appeals for the district in which he resides or has his principal place of business for a judicial review of such denial or revocation, pursuant to sections 701–706 of Title 5, United States Code.

(f) Licensees and holders of user permits shall make available for inspection at all reasonable times their records kept pursuant to this chapter or the regulations issued hereunder, and licensees and permittees shall submit to the Attorney General such reports and information with respect to such records and the contents thereof as he shall by regulations prescribe. The Attorney General may enter during business hours the premises (including places of storage) of any licensee or holder of a user permit, for the purpose of inspecting or examining (1) any records or documents required to be kept by such licensee or permittee, under the provisions of this chapter or regulations issued hereunder, and (2) any explosive materials kept or stored by such licensee or permittee at such premises. Upon the request of any State or any political subdivision thereof, the Attorney General may make available to such State or any political subdivision thereof, any information which he may obtain by reason of the provisions of this chapter with respect to the identification of persons within such State or political subdivision thereof, who have purchased or received explosive materials, together with a description of such explosive materials. The Attorney General may inspect the places of storage for explosive materials of an applicant for a limited permit or, at the time of renewal of such permit, a holder of a limited permit, only as provided in subsection (b)(4).

(g) Licenses and user permits issued under the provisions of subsection (b) of this section shall be kept posted and kept available for inspection on the premises covered by the license and permit.

(h) **(1)** If the Attorney General receives, from an employer, the name and other identifying information of a responsible person or an employee who will be authorized by the employer to possess explosive materials in the course of employment with the employer, the Attorney General shall determine whether the responsible person or employee is one of the persons described in any paragraph of section 842(i). In making the determination, the Attorney General may take into account a letter or document issued under paragraph (2).

(2) **(A)** If the Attorney General determines that the responsible person or the employee is not one of the persons described in any paragraph of section 842(i), the Attorney General shall notify the employer in writing or electronically of the determination and issue, to the responsible person or employee, a letter of clearance, which confirms the determination.

(B) If the Attorney General determines that the responsible person or employee is one of the persons described in any paragraph of section 842(i), the Attorney General shall notify the employer in writing or electronically of the determination and issue to the responsible person or the employee, as the case may be, a document that—

(i) confirms the determination;

(ii) explains the grounds for the determination;

(iii) provides information on how the disability may be relieved; and

(iv) explains how the determination may be appealed.

(i) Furnishing of Samples.

(1) In general. Licensed manufacturers and licensed importers and persons who manufacture or import explosive materials or ammonium nitrate shall, when required by letter issued by the Attorney General, furnish—

(A) samples of such explosive materials or ammonium nitrate;

(B) information on chemical composition of those products; and

(C) any other information that the Attorney General determines is relevant to the identification of the explosive materials or to identification of the ammonium nitrate.

(2) Reimbursement. The Attorney General shall, by regulation, authorize reimbursement of the fair market value of samples furnished pursuant to this subsection, as well as the reasonable costs of shipment.

(Added Pub. L. 91–452, title XI, Sec. 1102(a), Oct. 15, 1970, 84 Stat. 955; amended Pub. L. 107–296, title XI, Secs. 1112(e)(3), 1122(c)–(h), 1124, Nov. 25, 2002, 116 Stat. 2276, 2281, 2282, 2285.)

§ 844. Penalties

(a) Any person who—

(1) violates any of subsections (a) through (i) or (l) through (o) of section 842 shall be fined under this title, imprisoned for not more than 10 years, or both; and

(2) violates subsection (p)(2) of section 842, shall be fined under this title, imprisoned not more than 20 years, or both.

(b) Any person who violates any other provision of section 842 of this chapter shall be fined under this title or imprisoned not more than one year, or both.

(c) (1) Any explosive materials involved or used or intended to be used in any violation of the provisions of this chapter or any other rule or regulation promulgated thereunder or any violation of any criminal law of the United States shall be subject to seizure and forfeiture, and all provisions of the Internal Revenue Code of 1986 relating to the seizure, forfeiture, and disposition of firearms, as defined in section 5845(a) of that Code, shall, so far as applicable, extend to seizures and forfeitures under the provisions of this chapter.

(2) Notwithstanding paragraph (1), in the case of the seizure of any explosive materials for any offense for which the materials would be subject to forfeiture in which it would be impracticable or unsafe to remove the materials to a place of storage or would be unsafe to store them, the seizing officer may destroy the explosive materials forthwith. Any destruction under this paragraph shall be in the presence of at least 1 credible witness. The seizing officer shall make a report of the seizure and take samples as the Attorney General may by regulation prescribe.

(3) Within 60 days after any destruction made pursuant to paragraph (2), the owner of (including any person having an interest in) the property so destroyed may make application to the Attorney General for reimbursement of the value of the property. If the claimant establishes to the satisfaction of the Attorney General that—

(A) the property has not been used or involved in a violation of law; or

(B) any unlawful involvement or use of the property was without the claimant's knowledge, consent, or willful blindness, the Attorney General shall make an allowance to the claimant not exceeding the value of the property destroyed.

(d) Whoever transports or receives, or attempts to transport or receive, in interstate or foreign commerce any explosive with the knowledge or intent that it will be used to kill, injure, or intimidate any individual or unlawfully to damage or destroy any building, vehicle, or other real or personal property, shall be imprisoned for not more than ten years, or fined under this title, or both; and if personal injury results to any person, including any public safety officer performing duties as a direct or proximate result of conduct prohibited by this subsection, shall be imprisoned for not more than twenty years or fined under this title, or both; and if death results to any person, including any public safety officer performing duties as a direct or proximate result of conduct prohibited by this subsection, shall be subject to imprisonment for any term of years, or to the death penalty or to life imprisonment.

(e) Whoever, through the use of the mail, telephone, telegraph, or other instrument of interstate or foreign commerce, or in or affecting interstate or foreign commerce, willfully makes any threat, or maliciously conveys false information knowing the same to be false, concerning an attempt or alleged attempt being made, or to be made, to kill, injure, or intimidate any individual or unlawfully to damage or destroy any building, vehicle, or other real or personal property by means of fire or an explosive shall be imprisoned for not more than 10 years or fined under this title, or both.

(f) (1) Whoever maliciously damages or destroys, or attempts to damage or destroy, by means of fire or an explosive, any building, vehicle, or other personal or real property in whole or in part owned or possessed by, or leased to, the United States, or any department or agency thereof, or any institution or organization receiving Federal financial assistance, shall be imprisoned for not less than 5 years and not more than 20 years, fined under this title, or both.

(2) Whoever engages in conduct prohibited by this subsection, and as a result of such conduct, directly or proximately causes personal injury or creates a substantial risk of injury to any person, including any public safety officer performing duties, shall be imprisoned for not less than 7 years and not more than 40 years, fined under this title, or both.

(3) Whoever engages in conduct prohibited by this subsection, and as a result of such conduct directly or proximately causes the death of any person, including any public safety officer performing duties, shall be subject to the death penalty, or imprisoned for not less than 20 years or for life, fined under this title, or both.

(g) (1) Except as provided in paragraph (2), whoever possesses an explosive in an airport that is subject to the regulatory authority of the Federal Aviation Administration, or in any building in whole or in part owned, possessed, or used by, or leased to, the United States or any department or agency thereof, except with the written consent of the agency, department, or other person responsible for the management of such building or airport, shall be imprisoned for not more than five years, or fined under this title, or both.

(2) The provisions of this subsection shall not be applicable to—

(A) the possession of ammunition (as that term is defined in regulations issued pursuant to this chapter) in an airport that is subject to the regulatory authority of the Federal Aviation Administration if such ammunition is either in checked baggage or in a closed container; or

(B) the possession of an explosive in an airport if the packaging and transportation of such explosive is exempt from, or subject to and in accordance with, regulations of the Pipeline and Hazardous Materials Safety Administration for the handling of hazardous materials pursuant to chapter 51 of title 49.

(h) Whoever—

(1) uses fire or an explosive to commit any felony which may be prosecuted in a court of the United States, or

(2) carries an explosive during the commission of any felony which may be prosecuted in a court of the United States, including a felony which provides for an enhanced punishment if committed by the use of a deadly or dangerous weapon or device

shall, in addition to the punishment provided for such felony, be sentenced to imprisonment for 10 years. In the case of a second or subsequent conviction under this subsection, such person shall be sentenced to imprisonment for 20 years. Notwithstanding any other provision of law, the court shall not place on probation or suspend the sentence of any person convicted of a violation of this subsection, nor shall the term of imprisonment imposed under this subsection run concurrently with any other term of imprisonment including that imposed for the felony in which the explosive was used or carried.

(i) Whoever maliciously damages or destroys, or attempts to damage or destroy, by means of fire or an explosive, any building, vehicle, or other real or personal property used in interstate or foreign commerce or in any activity affecting interstate or foreign commerce shall be imprisoned for not less than 5 years and not more than 20 years, fined under this title, or both; and if personal injury results to any person, including any public safety officer performing duties as a direct or proximate result of conduct prohibited by this subsection, shall be imprisoned for not less than 7 years and not more than 40 years, fined under this title, or both; and if death results to any person, including any public safety officer performing duties as a direct or proximate result of conduct prohibited by this subsection, shall also be subject to imprisonment for any term of years, or to the death penalty or to life imprisonment.

(j) For the purposes of subsections (d), (e), (f), (g), (h), and (i) of this section and section 842(p), the term "explosive" means gunpowders, powders used for blasting, all forms of high explosives, blasting materials, fuzes (other than electric circuit breakers), detonators, and other detonating agents, smokeless powders, other explosive or incendiary devices within the meaning of paragraph (5) of section 232 of this title, and any chemical compounds, mechanical mixture, or device that contains any oxidizing and combustible units, or other ingredients, in such proportions, quantities, or packing that ignition by fire, by friction, by concussion, by percussion, or by detonation of the compound, mixture, or device or any part thereof may cause an explosion.

(k) A person who steals any explosives materials which are moving as, or are a part of, or which have moved in, interstate or foreign commerce shall be imprisoned for not more than 10 years, fined under this title, or both.

(l) A person who steals any explosive material from a licensed importer, licensed manufacturer, or licensed dealer, or from any permittee shall be fined under this title, imprisoned not more than 10 years, or both.

(m) A person who conspires to commit an offense under subsection (h) shall be imprisoned for any term of years not exceeding 20, fined under this title, or both.

(n) Except as otherwise provided in this section, a person who conspires to commit any offense defined in this chapter shall be subject to the same penalties (other than the penalty of death) as the penalties prescribed for the offense the commission of which was the object of the conspiracy.

(o) Whoever knowingly transfers any explosive materials, knowing or having reasonable cause to believe that such explosive materials will be used to commit a crime of violence (as defined in section 924(c)(3)) or drug trafficking crime (as defined in section 924(c)(2)) shall be subject to the same penalties as may be imposed under subsection (h) for a first conviction for the use or carrying of an explosive material.

(p) Theft Reporting Requirement.

(1) In general. A holder of a license or permit who knows that explosive materials have been stolen from that licensee or permittee, shall report the theft to the Attorney General not later than 24 hours after the discovery of the theft.

(2) Penalty. A holder of a license or permit who does not report a theft in accordance with paragraph (1), shall be fined not more than $10,000, imprisoned not more than 5 years, or both.

(Added Pub. L. 91–452, title XI, Sec. 1102(a), Oct. 15, 1970, 84 Stat. 956; amended Pub. L. 97–298, Sec. 2, Oct. 12, 1982, 96 Stat. 1319; Pub. L. 98–473, title II, Sec. 1014, Oct. 12, 1984, 98 Stat. 2142; Pub. L. 99–514, Sec. 2, Oct. 22, 1986, 100 Stat. 2095; Pub. L. 100–690, title VI, Sec. 6474(a), (b), Nov. 18, 1988, 102 Stat. 4379; Pub. L. 101–647, title XXXV, Sec. 3522, Nov. 29, 1990, 104 Stat. 4924; Pub. L. 103–272, Sec. 5(e)(7), July 5, 1994, 108 Stat. 1374; Pub. L. 103–322, title VI, Sec. 60003(a)(3), title XI, Secs. 110504(b), 110509, 110515(b), 110518(b), title XXXII, Secs. 320106, 320917(a), title XXXIII, Sec. 330016(1)(H), (K), (L), (N), Sept. 13, 1994, 108 Stat. 1969, 2016, 2018, 2020, 2111, 2129, 2147, 2148; Pub. L. 104–132, title VI, Sec. 604, title VII, Secs. 701, 706, 708(a), (c)(3), 724, Apr. 24, 1996, 110 Stat. 1289, 1291, 1295–1297, 1300; Pub. L. 104–294, title VI, Sec. 603(a), Oct. 11, 1996, 110 Stat. 3503; Pub. L. 106–54, Sec. 2(b), Aug. 17, 1999, 113 Stat. 399; Pub. L. 107–296, title XI, Secs. 1112(e)(3), 1125, 1127, Nov. 25, 2002, 116 Stat. 2276, 2285; Pub. L. 108–426, Sec. 2(c)(6), Nov. 30, 2004, 118 Stat. 2424.)

§845. Exceptions; Relief From Disabilities

(a) Except in the case of subsection (l), (m), (n), or (o) of section 842 and subsections (d), (e), (f), (g), (h), and (i) of section 844 of this title, this chapter shall not apply to:

(1) aspects of the transportation of explosive materials via railroad, water, highway, or air that pertain to safety, including security, and are regulated by the Department of Transportation or the Department of Homeland Security;

(2) the use of explosive materials in medicines and medicinal agents in the forms prescribed by the official United States Pharmacopeia, or the National Formulary;

(3) the transportation, shipment, receipt, or importation of explosive materials for delivery to any agency of the United States or to any State or political subdivision thereof;

(4) small arms ammunition and components thereof;

(5) commercially manufactured black powder in quantities not to exceed fifty pounds, percussion caps, safety and pyrotechnic fuses, quills, quick and slow matches, and friction primers, intended to be used solely for sporting, recreational, or cultural purposes in antique firearms as defined in section 921(a)(16) of title 18 of the United States Code, or in antique devices as exempted from the term "destructive device" in section 921(a)(4) of title 18 of the United States Code;

(6) the manufacture under the regulation of the military department of the United States of explosive materials for, or their distribution to or storage or possession by the military or naval services or other agencies of the United States; or to arsenals, navy yards, depots, or other establishments owned by, or operated by or on behalf of, the United States and

(7) the transportation, shipment, receipt, or importation of display fireworks materials for delivery to a federally recognized Indian tribe or tribal agency.

(b) (1) A person who is prohibited from shipping, transporting, receiving, or possessing any explosive under section 842(i) may apply to the Attorney General for relief from such prohibition.

(2) The Attorney General may grant the relief requested under paragraph (1) if the Attorney General determines that the circumstances regarding the applicability of section 842(i), and the applicant's record and reputation, are such that the applicant will not be likely to act in a manner dangerous to public safety and that the granting of such relief is not contrary to the public interest.

(3) A licensee or permittee who applies for relief, under this subsection, from the disabilities incurred under this chapter as a result of an indictment for or conviction of a crime punishable by imprisonment for a term exceeding 1 year shall not be barred by such disability from further operations under the license or permit pending final action on an application for relief filed pursuant to this section.

(c) It is an affirmative defense against any proceeding involving subsections (l) through (o) of section 842 if the proponent proves by a preponderance of the evidence that the plastic explosive—

(1) consisted of a small amount of plastic explosive intended for and utilized solely in lawful—

(A) research, development, or testing of new or modified explosive materials;

(B) training in explosives detection or development or testing of explosives detection equipment; or

(C) forensic science purposes; or

(2) was plastic explosive that, within 3 years after the date of enactment of the Antiterrorism and Effective Death Penalty Act of 1996 [enacted April 24, 1996], will be or is incorporated in a military device within the territory of the United States and remains an integral part of such military device, or is intended to be, or is incorporated in, and remains an integral part of a military device that is intended to become, or has become, the property of any agency of the United States performing military or police functions (including any military reserve component) or the National Guard of any State, wherever such device is located.

(3) For purposes of this subsection, the term "military device" includes, but is not restricted to, shells, bombs, projectiles, mines, missiles, rockets, shaped charges, grenades, perforators, and similar devices lawfully manufactured exclusively for military or police purposes.

§846. Additional Powers of the Attorney General

(a) The Attorney General is authorized to inspect the site of any accident, or fire, in which there is reason to believe that explosive materials were involved, in order that if any such incident has been brought about by accidental means, precautions may be taken to prevent similar accidents from occurring. In order to carry out the purpose of this subsection, the Attorney General is authorized to enter into or upon any property where explosive materials have been used, are suspected of having been used, or have been found in an otherwise unauthorized location. Nothing in this chapter shall be construed as modifying or otherwise affecting in any way the investigative authority of any other Federal agency. In addition to any other investigatory authority they have with respect to violations of provisions of this chapter, the Federal Bureau of Investigation, together with the Bureau of Alcohol, Tobacco, Firearms, and Explosives, shall have authority to conduct investigations with respect to violations of subsection (d), (e), (f), (g), (h), or (i) of section 844 of this title.

(b) The Attorney General is authorized to establish a national repository of information on incidents involving arson and the suspected criminal misuse of explosives. All Federal agencies having information concerning such incidents shall report the information to the Attorney General pursuant to such regulations as deemed necessary to carry out the provisions of this subsection. The repository shall also contain information on incidents voluntarily reported to the Attorney General by State and local authorities.

(Added Pub. L. 91–452, title XI, Sec. 1102(a), Oct. 15, 1970, 84 Stat. 959; amended Pub. L. 104–208, div. A, title I, Sec. 101(f) [title VI, Sec. 654(a)], Sept. 30, 1996, 110 Stat. 3009–314, 3009–369; Pub. L. 107–296, title XI, Sec. 1112(e)(2), (3), Nov. 25, 2002, 116 Stat. 2276.)

§847. Rules and Regulations

The administration of this chapter shall be vested in the Attorney General. The Attorney General may prescribe such rules and regulations as he deems reasonably necessary to carry out the provisions of this chapter. The Attorney General shall give reasonable public notice, and afford to interested parties opportunity for hearing, prior to prescribing such rules and regulations.

(Added Pub. L. 91–452, title XI, Sec. 1102(a), Oct. 15, 1970, 84 Stat. 959; amended Pub. L. 107–296, title XI, Sec. 1112(e)(3), Nov. 25, 2002, 116 Stat. 2276.)

§848. Effect on State Law

No provision of this chapter shall be construed as indicating an intent on the part of the Congress to occupy the field in which such provision operates to the exclusion of the law of any State on the same subject matter, unless there is a direct and positive conflict between such provision and the law of the State so that the two cannot be reconciled or consistently stand together.

(Added Pub. L. 91–452, title XI, Sec. 1102(a), Oct. 15, 1970, 84 Stat. 959.)

Title 27 [Code of Federal Regulations]
Part 555—Commerce in Explosives

Section Contents

Subpart H—Exemptions

Subpart I—Unlawful Acts, Penalties, Seizures, and Forfeitures

Subpart J—Marking of Plastic Explosives

Subpart K—Storage

Authority: 18 U.S.C. 847.

Source: T.D. ATF–87, 46 FR 40384, Aug. 7, 1981, unless otherwise noted. Redesignated by T.D. ATF–487, 68 FR 3748, Jan. 24, 2003.

Editorial Note: Nomenclature changes to part 555 appear at 68 FR 3748, Jan. 24, 2003. Also, there are several name changes for titles referenced in the regulations, 27 CFR Part 555. The title "Chief, Firearms and Explosives Licensing Center" has been changed to "Chief, Federal Explosives Licensing Center". The title "Regional director (compliance)" has been changed to "Director, Industry Operations" within the respective ATF Field Divisions. All regulations and rulings contained within this publication were current as of 3/13/12. For more recent information, please refer to the ATF website at www.atf.gov/.

Subpart A—Introduction

§ 555.1 Scope of regulations.

(a) In general. The regulations contained in this part relate to commerce in explosives and implement Title XI, Regulation of Explosives (18 U.S.C. Chapter 40; 84 Stat. 952), of the Organized Crime Control Act of 1970 (84 Stat. 922), Pub. L. 103–322 (108 Stat. 1796), Pub. L. 104–132 (110 Stat. 1214), and Pub. L. 107–296 (116 Stat. 2135).

(b) Procedural and substantive requirements. This part contains the procedural and substantive requirements relative to:

(1) The interstate or foreign commerce in explosive materials;

(2) The licensing of manufacturers and importers of, and dealers in, explosive materials;

(3) The issuance of permits;

(4) The conduct of business by licensees and operations by permittees;

(5) The storage of explosive materials;

(6) The records and reports required of licensees and permittees;

(7) Relief from disabilities under this part;

(8) Exemptions, unlawful acts, penalties, seizures, and forfeitures; and

(9) The marking of plastic explosives.

[T.D. ATF–87, 46 FR 40384, Aug. 7, 1981, as amended by T.D. ATF–363, 60 FR 17449, Apr. 6, 1995; T.D. ATF–387, 62 FR 8376, Feb. 25, 1997; ATF No. 1, 68 FR 13780, Mar. 20, 2003]

§ 555.2 Relation to other provisions of law.

The provisions in this part are in addition to, and are not in lieu of, any other provision of law, or regulations, respecting commerce in explosive materials. For regulations applicable to commerce in firearms and ammunition, see Part 478 of this chapter. For regulations applicable to traffic in machine guns, destructive devices, and certain other firearms, see Part 479 of this chapter. For statutes applicable to the registration and licensing of persons engaged in the business of manufacturing, importing or exporting arms, ammunition, or implements of war, see section 38 of the Arms Export Control Act (22 U.S.C. 2778), and regulations of Part 447 of this chapter and in Parts 121 through 128 of Title 22, Code of Federal Regulations. For statutes applicable to nonmailable materials, see 18 U.S.C. 1716 and implementing regulations. For statutes applicable to water quality standards, see 33 U.S.C. 1341.

Subpart B—Definitions

§ 555.11 Meaning of terms.

When used in this part, terms are defined as follows in this section. Words in the plural form include the singular, and vice versa, and words indicating the masculine gender include the feminine. The terms "includes" and "including" do not exclude other things not named which are in the same general class or are otherwise within the scope of the term defined.

Act. 18 U.S.C. Chapter 40.

Adjudicated as a mental defective. (a) A determination by a court, board, commission, or other lawful authority that a person, as a result of marked subnormal intelligence, or mental illness, incompetency, condition, or disease:

(1) Is a danger to himself or to others; or

(2) Lacks the mental capacity to contract or manage his own affairs.

(b) The term will include—

(1) A finding of insanity by a court in a criminal case; and

(2) Those persons found incompetent to stand trial or found not guilty by reason of lack of mental responsibility by any court or pursuant to articles 50a and 76b of the Uniform Code of Military Justice, 10 U.S.C. 850a, 876b.

Alien. Any person who is not a citizen or national of the United States.

Ammunition. Small arms ammunition or cartridge cases, primers, bullets, or smokeless propellants designed for use in small arms, including percussion caps, and 3/32 inch and other external burning pyrotechnic hobby fuses. The term does not include black powder.

Appropriate identifying information. The term means, in relation to an individual:

(a) The full name, date of birth, place of birth, sex, race, street address, State of residence, telephone numbers (home and work), country or countries of citizenship, and position at the employer's business or operations of responsible persons and employees authorized to possess explosive materials;

(b) The business name, address, and license or permit number with which the responsible person or employee is affiliated;

(c) If an alien, INS-issued alien number or admission number; and

(d) Social security number, as optional information (this information is not required but is helpful in avoiding misidentification when a background check is conducted).

Approved storage facility. A place where explosive materials are stored, consisting of one or more approved magazines, conforming to the requirements of this part and covered by a license or permit issued under this part.

Articles pyrotechnic. Pyrotechnic devices for professional use similar to consumer fireworks in chemical composition and construction but not intended for consumer use. Such articles meeting the weight limits for consumer fireworks but not labeled as such and classified by U.S. Department of Transportation regulations in 49 CFR 172.101 as UN0431 or UN0432.

Artificial barricade. An artificial mound or revetted wall of earth of a minimum thickness of three feet, or any other approved barricade that offers equivalent protection.

ATF. (a) Prior to January 24, 2003. The Bureau of Alcohol, Tobacco and Firearms, Department of the Treasury, Washington, DC.

(b) On and after January 24, 2003. The Bureau of Alcohol, Tobacco, Firearms and Explosives, Department of Justice, Washington, DC.

ATF officer. (a) Prior to January 24, 2003. An officer or employee of the Bureau of Alcohol, Tobacco and Firearms (ATF) authorized to perform any function relating to the administration or enforcement of this part.

(b) On and after January 24, 2003. An officer or employee of the Bureau of Alcohol, Tobacco, Firearms and Explosives (ATF) authorized to perform any function relating to the administration or enforcement of this part.

Authority having jurisdiction for fire safety. The fire department having jurisdiction over sites where explosives are manufactured or stored.

Barricaded. The effective screening of a magazine containing explosive materials from another magazine, a building, a railway, or a highway, either by a natural barricade or by an artificial barricade. To be properly barricaded, a straight line from the top of any sidewall of the magazine containing explosive materials to the eave line of any other magazine or building, or to a point 12 feet above the center of a railway or highway, will pass through the natural or artificial barricade.

Blasting agent. Any material or mixture, consisting of fuel and oxidizer, that is intended for blasting and not otherwise defined as an explosive; if the finished product, as mixed for use or shipment, cannot be detonated by means of a number 8 test blasting cap when unconfined. A number 8 test blasting cap is one containing 2 grams of a mixture of 80 percent mercury fulminate and 20 percent potassium chlorate, or a blasting cap of equivalent strength. An equivalent strength cap comprises 0.40–0.45 grams of PETN base charge pressed in an aluminum shell with bottom thickness not to exceed to 0.03 of an inch, to a specific gravity of not less than 1.4 g/cc., and primed with standard weights of primer depending on the manufacturer.

Bulk salutes. Salute components prior to final assembly into aerial shells, and finished salute shells held separately prior to being packed with other types of display fireworks.

Bullet-sensitive explosive materials. Explosive materials that can be exploded by 150-grain M2 ball ammunition having a nominal muzzle velocity of 2700 fps (824 mps) when fired from a .30 caliber rifle at a distance of 100 ft (30.5 m), measured perpendicular. The test material is at a temperature of 70 to 75 ° F (21 to 24 ° C) and is placed against a ½ " (12.4 mm) steel backing plate.

Bureau. (a) Prior to January 24, 2003. The Bureau of Alcohol, Tobacco and Firearms, Department of the Treasury, Washington, DC.

(b) On and after January 24, 2003. The Bureau of Alcohol, Tobacco, Firearms and Explosives, Department of Justice, Washington, DC.

Business premises. When used with respect to a manufacturer, importer, or dealer, the property on which explosive materials are manufactured, imported, stored or distributed. The premises include the property where the records of a manufacturer, importer, or dealer are kept if different than the premises where explosive materials are manufactured, imported, stored or distributed. When used with respect to a user of explosive materials, the property on which the explosive materials are received or stored. The premises includes the property where the records of the users are kept if different than the premises where explosive materials are received or stored.

Chief, Federal Explosives Licensing Center. The ATF official responsible for the issuance and renewal of licenses and permits under this part.

Committed to a mental institution. A formal commitment of a person to a mental institution by a court, board, commission, or other lawful authority. The term includes a commitment to a mental institution involuntarily. The term includes commitment for mental defectiveness or mental illness. It also includes commitments for other reasons, such as for drug use. The term does not include a person in a mental institution for observation or a voluntary admission to a mental institution.

Common or contract carrier. Any individual or organization engaged in the business of transporting passengers or goods.

Consumer fireworks. Any small firework device designed to produce visible effects by combustion and which must comply with the construction, chemical composition, and labeling regulations of the U.S. Consumer Product Safety Commission, as set forth in title 16, Code of Federal Regulations, parts 1500 and 1507. Some small devices designed to produce audible effects are included, such as whistling devices, ground devices containing 50 mg or less of explosive materials, and aerial devices containing 130 mg or less of explosive materials. Consumer fireworks are classified as fireworks UN0336, and UN0337 by the U.S. Department of Transportation at 49 CFR 172.101. This term does not include fused set pieces containing components which together exceed 50 mg of salute powder.

Controlled substance. A drug or other substance, or immediate precursor, as defined in section 102 of the Controlled Substances Act, 21 U.S.C. 802. The term includes, but is not limited to, marijuana, depressants, stimulants, and narcotic drugs. The term does not include distilled spirits, wine, malt beverages, or tobacco, as those terms are defined or used in Subtitle E of the Internal Revenue Code of 1986, as amended.

Crime punishable by imprisonment for a term exceeding one year. Any offense for which the maximum penalty, whether or not imposed, is capital punishment or imprisonment in excess of one year. The term does not include (a) any Federal or State offenses pertaining to antitrust violations, unfair trade practices, restraints of trade, or (b) any State offense (other than one involving a firearm or explosive) classified by the laws of the State as a misdemeanor and punishable by a term of imprisonment of two years or less.

Customs officer. Any officer of the Customs Service or any commissioned, warrant, or petty officer of the Coast Guard, or any agent or other person authorized to perform the duties of an officer of the Customs Service.

Dealer. Any person engaged in the business of distributing explosive materials at wholesale or retail.

Detonator. Any device containing a detonating charge that is used for initiating detonation in an explosive. The term includes, but is not limited to, electric blasting caps of instantaneous and delay types, blasting caps for use with safety fuses, detonating cord delay connectors, and nonelectric instantaneous and delay blasting caps.

Director. (a) Prior to January 24, 2003. The Director, Bureau of Alcohol, Tobacco and Firearms, Department of the Treasury, Washington, DC. (b) On and after January 24, 2003. The Director, Bureau of Alcohol, Tobacco, Firearms and Explosives, Department of Justice, Washington, DC.

Discharged under dishonorable conditions. Separation from the U.S. Armed Forces resulting from a dishonorable discharge or dismissal adjudged by general court-martial. The term does not include any separation from the Armed Forces resulting from any other discharge, e.g., a bad conduct discharge.

Display fireworks. Large fireworks designed primarily to produce visible or audible effects by combustion, deflagration, or detonation. This term includes, but is not limited to, salutes containing more than 2 grains (130 mg) of explosive materials, aerial shells containing more than 40 grams of pyrotechnic compositions, and other display pieces which exceed the limits of explosive materials for classification as "consumer fireworks." Display fireworks are classified as fireworks UN0333, UN0334 or UN0335 by the U.S. Department of Transportation at 49 CFR 172.101. This term also includes fused setpieces containing components which together exceed 50 mg of salute powder.

Distribute. To sell, issue, give, transfer, or otherwise dispose of. The term does not include a mere change of possession from a person to his agent or employee in connection with the agency or employment.

Executed under penalties of perjury. Signed with the required declaration under the penalties of perjury as provided on or with respect to the return, form, or other document or, where no form of declaration is required, with the declaration:

"I declare under the penalties of perjury that this—(insert type of document, such as, statement, application, request, certificate), including the documents submitted in support thereof, has been examined by me and, to the best of my knowledge and belief, is true, correct, and complete".

Explosive actuated device. Any tool or special mechanized device which is actuated by explosives, but not a propellent actuated device.

Explosive materials. Explosives, blasting agents, water gels and detonators. Explosive materials include, but are not limited to, all items in the "List of Explosive Materials" provided for in § 555.23.

Explosives. Any chemical compound, mixture, or device, the primary or common purpose of which is to function by explosion. The term includes, but is not limited to, dynamite and other high explosives, black powder, pellet powder, initiating explosives, detonators, safety fuses, squibs, detonating cord, igniter cord, and igniters.

Fireworks. Any composition or device designed to produce a visible or an audible effect by combustion, deflagration, or detonation, and which meets the definition of "consumer fireworks" or "display fireworks" as defined by this section.

Fireworks mixing building. Any building or area used for mixing and blending pyrotechnic compositions except wet sparkler mix.

Fireworks nonprocess building. Any office building or other building or area in a fireworks plant where no fireworks, pyrotechnic compositions or explosive materials are processed or stored.

Fireworks plant. All land and buildings thereon used for or in connection with the assembly or processing of fireworks, including warehouses used with or in connection with fireworks plant operations.

Fireworks plant warehouse. Any building or structure used exclusively for the storage of materials which are neither explosive materials nor pyrotechnic compositions used to manufacture or assemble fireworks.

Fireworks process building. Any mixing building; any building in which pyrotechnic compositions or explosive materials is pressed or otherwise prepared for finished and assembly; or any finishing or assembly building.

Fireworks shipping building. A building used for the packing of assorted display fireworks into shipping cartons for individual public displays and for the loading of packaged displays for shipment to purchasers.

Flash powder. An explosive material intended to produce an audible report and a flash of light when ignited which includes but is not limited to oxidizers such as potassium chlorate or potassium perchlorate, and fuels such as sulfur or aluminum powder.

Fugitive from justice. Any person who has fled from the jurisdiction of any court of record to avoid prosecution for any crime or to avoid giving testimony in any criminal proceeding.

The term also includes any person who has been convicted of any crime and has fled to avoid imprisonment.

Hardwood. Oak, maple, ash, hickory, or other hard wood, free from loose knots, spaces, or similar defects.

Highway. Any public street, public alley, or public road, including a privately financed, constructed, or maintained road that is regularly and openly traveled by the general public.

Identification document. A document containing the name, residence address, date of birth, and photograph of the holder and which was made or issued by or under the authority of the United States Government, a State, political subdivision of a State, a foreign government, a political subdivision of a foreign government, an international governmental or an international quasi-governmental organization which, when completed with information concerning a particular individual, is of a type intended or commonly accepted for the purpose of identification of individuals.

Importer. Any person engaged in the business of importing or bringing explosive materials into the United States for purposes of sale or distribution.

Indictment. Includes an indictment or information in any court under which a crime punishable by imprisonment for a term exceeding one year may be prosecuted.

Inhabited building. Any building regularly occupied in whole or in part as a habitation for human beings, or any church, school-house, railroad station, store, or other structure where people are accustomed to assemble, except any building occupied in connection with the manufacture, transportation, storage, or use of explosive materials.

Interstate or foreign commerce. Commerce between any place in a State and any place outside of that State, or within any possession of the United States or the District of Columbia, and commerce between places within the same State but through any place outside of that State.

Licensed dealer. A dealer licensed under this part.

Licensed importer. An importer licensed under this part.

Licensed manufacturer. A manufacturer licensed under this part to engage in the business of manufacturing explosive materials for purposes of sale or distribution or for his own use.

Licensee. Any importer, manufacturer, or dealer licensed under this part.

Limited permit. A permit issued to a person authorizing him to receive for his use explosive materials from a licensee or permittee in his state of residence on no more than 6 occasions during the 12-month period in which the permit is valid. A limited permit does not authorize the receipt or transportation of explosive materials in interstate or foreign commerce.

Magazine. Any building or structure, other than an explosives manufacturing building, used for storage of explosive materials.

Manufacturer. Any person engaged in the business of manufacturing explosive materials for purposes of sale or distribution or for his own use.

Mass detonation (mass explosion). Explosive materials mass detonate (mass explode) when a unit or any part of a larger quantity of explosive material explodes and causes all or a substantial part of the remaining material to detonate or explode.

Mental institution. Includes mental health facilities, mental hospitals, sanitariums, psychiatric facilities, and other facilities that provide diagnoses by licensed professionals of mental retardation or mental illness, including a psychiatric ward in a general hospital.

Natural barricade. Natural features of the ground, such as hills, or timber of sufficient density that the surrounding exposures which require protection cannot be seen from the magazine when the trees are bare of leaves.

Number 8 test blasting cap. (See definition of "blasting agent.")

Permittee. Any user of explosives for a lawful purpose who has obtained either a user permit or a limited permit under this part.

Person. Any individual, corporation, company, association, firm, partnership, society, or joint stock company.

Plywood. Exterior, construction grade (laminated wood) plywood.

Propellant actuated device. (a) Any tool or special mechanized device or gas generator system that is actuated by a propellant or which releases and directs work through a propellant charge.

(b) The term does not include—

(1) Hobby rocket motors consisting of ammonium perchlorate composite propellant, black powder, or other similar low explosives, regardless of amount; and

(2) Rocket-motor reload kits that can be used to assemble hobby rocket motors containing ammonium perchlorate composite propellant, black powder, or other similar low explosives, regardless of amount.

Pyrotechnic compositions. A chemical mixture which, upon burning and without explosion, produces visible, brilliant displays, bright lights, or sounds.

Railway. Any steam, electric, or other railroad or railway which carries passengers for hire.

Region. A geographical region of the Bureau of Alcohol, Tobacco, Firearms and Explosives.

Regional director (compliance). (Changed to "Director, Industry Operations" within the respective ATF Field Divisions.) The principal regional official responsible for administering regulations in this part.

Renounced U.S. citizenship. (a) A person has renounced his U.S. citizenship if the person, having been a citizen of the United States, has renounced citizenship either—

(1) Before a diplomatic or consular officer of the United States in a foreign state pursuant to 8 U.S.C. 1481(a)(5); or

(2) Before an officer designated by the Attorney General when the United States is in a state of war pursuant to 8 U.S.C. 1481(a)(6).

(b) The term will not include any renunciation of citizenship that has been reversed as a result of administrative or judicial appeal.

Responsible person. An individual who has the power to direct the management and policies of the applicant pertaining to explosive materials. Generally, the term includes partners, sole proprietors, site managers, corporate officers and directors, and majority shareholders.

Salute. An aerial shell, classified as a display firework, that contains a charge of flash powder and is designed to produce a flash of light and a loud report as the pyrotechnic effect.

Screen barricade. Any barrier that will contain the embers and debris from a fire or deflagration in a process building, thus preventing propagation of fire to other buildings or areas. Such barriers shall be constructed of metal roofing, ¼ to ½" (6 to 13 mm) mesh screen, or equivalent material. The barrier extends from floor level to a height such that a straight line from the top of any side wall of the donor building to the eave line of any exposed building intercepts the screen at a point not less than 5 feet (1.5 m) from the top of the screen. The top 5 feet (1.5 m) of the screen is inclined towards the donor building at an angle of 30 to 45°.

Softwood. Fir, pine, or other soft wood, free from loose knots, spaces, or similar defects.

State. A State of the United States. The term includes the District of Columbia, the Commonwealth of Puerto Rico, and the possessions of the United States.

State of residence. The State in which an individual regularly resides or maintains his home. Temporary stay in a State does not make the State of temporary stay the State of residence.

Theatrical flash powder. Flash powder commercially manufactured in premeasured kits not exceeding 1 ounce and mixed immediately prior to use and intended for use in theatrical shows, stage plays, band concerts, magic acts, thrill shows, and clown acts in circuses.

Unlawful user of or addicted to any controlled substance. A person who uses a controlled substance and has lost the power of self-control with reference to the use of a controlled substance; and any person who is a current user of a controlled substance in a manner other than as prescribed by a licensed physician. Such use is not limited to the use of drugs on a particular day, or within a matter of days or weeks before possession of the explosive materials, but rather that the unlawful use has occurred recently enough to indicate that the individual is actively engaged in such conduct. A person may be an unlawful current user of a controlled substance even though the substance is not being used at the precise time the person seeks to acquire explosive materials or receives or possesses explosive materials. An inference of current use may be drawn from evidence of a recent use or possession of a controlled substance or a pattern of use or possession that reasonably covers the present time, e.g., a conviction for use or possession of a controlled substance within the past year; multiple arrests for such offenses within the past 5 years if the most recent arrest occurred within the past year; or persons found through a drug test to use a controlled substance unlawfully, provided that the test was administered within the past year. For a current or former member of the Armed Forces, an inference of current use may be drawn from recent disciplinary or other administrative action based on confirmed drug use, e.g., court-martial conviction, nonjudicial punishment, or an administrative discharge based on drug use or drug rehabilitation failure.

U.S.C. The United States Code.

User-limited permit. A user permit valid only for a single purchase transaction, a new permit being required for a subsequent purchase transaction.

User permit. A permit issued to a person authorizing him (a) to acquire for his own use explosive materials from a licensee in a State other than the State in which he resides or from a foreign country, and (b) to transport explosive materials in interstate or foreign commerce.

Water gels. Explosives or blasting agents that contain a substantial proportion of water.

(18 U.S.C. 847 (84 Stat. 959); 18 U.S.C. 926 (82 Stat. 1226)

[T.D. ATF–87, 46 FR 40384, Aug. 7, 1981]

Subpart C—Administrative and Miscellaneous Provisions

§ 555.21 Forms prescribed.

(a) The Director is authorized to prescribe all forms required by this part. All of the information called for in each form shall be furnished as indicated by the headings on the form and the instructions on or pertaining to the form. In addition, information called for in each form shall be furnished as required by this part.

(b) Requests for forms should be mailed to the ATF Distribution Center, 1519 Cabin Branch Drive, Landover, MD 20785.

[T.D. ATF–92, 46 FR 46916, Sept. 23, 1981, as amended by T.D. ATF–249, 52 FR 5961, Feb. 27, 1987; T.D. 372, 61 FR 20724, May 8, 1996; ATF–11F, 73 FR 57242, Oct. 2, 2008]

§ 555.22 Alternate methods or procedures; emergency variations from requirements.

(a) Alternate methods or procedures. The permittee or licensee, on specific approval by the Director as provided by this paragraph, may use an alternate method or procedure in lieu of a method or procedure specifically prescribed in this part. The Director may approve an alternate method or procedure, subject to stated conditions, when he finds that:

(1) Good cause is shown for the use of the alternate method or procedure;

(2) The alternate method or procedure is within the purpose of, and consistent with the effect intended by, the specifically prescribed method or procedure and that the alternate method or procedure is substantially equivalent to that specifically prescribed method or procedure; and

(3) The alternate method or procedure will not be contrary to any provision of law and will not result in an increase in cost to the Government or hinder the effective administration of this part.

Where the permittee or licensee desires to employ an alternate method or procedure, he shall submit a written application to the Director, Industry Operations, for transmittal to the Director. The application shall specifically describe the proposed alternate method or procedure and shall set forth the reasons for it. Alternate methods or procedures may not be employed until the application is approved by the Director. The permittee or licensee shall, during the period of authorization of an alternate method or procedure, comply with the terms of the approved application. Authorization of any alternate method or procedure may be withdrawn whenever, in the judgment of the Director, the effective administration of this part is hindered by the continuation of the authorization. As used in this paragraph, alternate methods or procedures include alternate construction or equipment.

(b) Emergency variations from requirements. The Director may approve construction, equipment, and methods of operation other than as specified in this part, where he finds that an emergency exists and the proposed variations from the specified requirements are necessary and the proposed variations:

(1) Will afford security and protection that are substantially equivalent to those prescribed in this part;

(2) Will not hinder the effective administration of this part; and

(3) Will not be contrary to any provisions of law.

Variations from requirements granted under this paragraph are conditioned on compliance with the procedures, conditions, and limitations set forth in the approval of the application. Failure to comply in good faith with the procedures, conditions, and limitations shall automatically terminate the authority for the variations and the licensee or permittee shall fully comply with the prescribed requirements of regulations from which the variations were authorized. Authority for any variation may be withdrawn whenever, in the judgment of the Director, the effective administration of this part is hindered by the continuation of the variation. Where the licensee or permittee desires to employ an emergency variation, he shall submit a written application to the Director, Industry Operations for transmittal to the Director. The application shall describe the proposed variation and set forth the reasons for it. Variations may not be employed until the application is approved, except when the emergency requires immediate action to correct a situation that is threatening to life or property. Corrective action may then be taken concurrent with the filing of the application and notification of the Director via telephone.

(c) Retention of approved variations. The licensee or permittee shall retain, as part of his records available for examination by ATF officers, any application approved by the Director under this section.

§ 555.23 List of explosive materials.

The Director shall compile a list of explosive materials, which shall be published and revised at least annually in the Federal Register. The "List of Explosive Materials" (ATF Publication 5400.8) is available at no cost upon request from the ATF Distribution Center (See § 555.21).

[T.D. ATF–290, 54 FR 53054, Dec. 27, 1989, as amended by T.D. ATF–446, 66 FR 16602, Mar. 27, 2001; ATF–11F, 73 FR 57242, Oct. 2, 2008]

§ 555.24 Right of entry and examination.

(a) Any ATF officer may enter during business hours the premises, including places of storage, of any licensee or holder of a user permit for the purpose of inspecting or examining any records or documents required to be kept under this part, and any facilities in which explosive materials are kept or stored.

(b) Any ATF officer may inspect the places of storage for explosive materials of an applicant for a limited permit or, in the case of a holder of a limited permit, at the time of renewal of such permit.

(c) The provisions of paragraph (b) of this section do not apply to an applicant for the renewal of a limited permit if an ATF officer has, within the preceding 3 years, verified by inspection that the applicant's place of storage for explosive materials meets the requirements of subpart K of this part.

[ATF No. 1, 68 FR 13781, Mar. 20, 2003]

§ 555.25 Disclosure of information.

Upon receipt of written request from any State or any political subdivision of a State, the Director, Industry Operations may make available to the State or political subdivision any information which the Director, Industry Operations may obtain under the Act with respect to the identification of persons within the State or political subdivision, who have purchased or received explosive materials, together with a description of the explosive materials.

§ 555.26 Prohibited shipment, transportation, receipt, possession, or distribution of explosive materials.

(a) General. No person, other than a licensee or permittee knowingly may transport, ship, cause to be transported, or receive any explosive materials: Provided, that the provisions of this paragraph (a) do not apply to the lawful purchase by a nonlicensee or nonpermittee of commercially manufactured black powder in quantities not to exceed 50 pounds, if the black powder is intended to be used solely for sporting, recreational, or cultural purposes in antique firearms as defined in 18 U.S.C. 921(a)(16), or in antique devices as exempted from the term "destructive device" in 18 U.S.C. 921(a)(4).

(b) Holders of a limited permit. No person who is a holder of a limited permit may—

(1) Transport, ship, cause to be transported, or receive in interstate or foreign commerce any explosive materials;

(2) Receive explosive materials from a licensee or permittee, whose premises are located outside the State of residence of the limited permit holder; or

(3) Receive explosive materials on more than 6 separate occasions, during the period of the permit, from one or more licensees or permittees whose premises are located within the State of residence of the limited permit holder. (See § 555.105(b) for the definition of "6 separate occasions.")

(c) Possession by prohibited persons. No person may ship or transport any explosive material in or affecting interstate or foreign commerce or receive or possess any explosive materials which have been shipped or transported in or affecting interstate or foreign commerce who:

(1) Is under indictment or information for, or who has been convicted in any court of, a crime punishable by imprisonment for a term exceeding one year;

(2) Is a fugitive from justice;

(3) Is an unlawful user of or addicted to any controlled substance (as defined in section 102 of the Controlled Substances Act (21 U.S.C. 802) and § 555.11);

(4) Has been adjudicated as a mental defective or has been committed to a mental institution;

(5) Is an alien, other than an alien who—

(i) Is lawfully admitted for permanent residence (as that term is defined in section 101(a)(20) of the Immigration and Nationality Act (8 U.S.C. 1101)); or

(ii) Is in lawful nonimmigrant status, is a refugee admitted under section 207 of the Immigration and Nationality Act (8 U.S.C. 1157), or is in asylum status under section 208 of the Immigration and Nationality Act (8 U.S.C. 1158), and—

(A) Is a foreign law enforcement officer of a friendly foreign government, as determined by the Attorney General in consultation with the Secretary of State, entering the United States on official law enforcement business, and the shipping, transporting, possession, or receipt of explosive materials is in furtherance of this official law enforcement business;

(B) Is a person having the power to direct or cause the direction of the management and policies of a corporation, partnership, or association licensed pursuant to section 843(a) of the Act, and the shipping, transporting, possession, or receipt of explosive materials is in furtherance of such power;

(C) Is a member of a North Atlantic Treaty Organization (NATO) or other friendly foreign military force, as determined by the Attorney General in consultation with the Secretary of Defense, (whether or not admitted in a nonimmigrant status) who is present in the United States under military orders for training or other military purpose authorized by the United States, and the shipping, transporting, possession, or receipt of explosive materials is in furtherance of the military purpose; or

(D) Is lawfully present in the United States in cooperation with the Director of Central Intelligence, and the shipment, transportation, receipt, or possession of the explosive materials is in furtherance of such cooperation;

(6) Has been discharged from the armed forces under dishonorable conditions; or

(7) Having been a citizen of the United States, has renounced citizenship.

(d) Distribution to prohibited persons. No person may knowingly distribute explosive materials to any individual who:

(1) Is under twenty-one years of age;

(2) Is under indictment or information for, or who has been convicted in any court of, a crime punishable by imprisonment for a term exceeding one year;

(3) Is a fugitive from justice;

(4) Is an unlawful user of or addicted to any controlled substance (as defined in section 102 of the Controlled Substances Act (21 U.S.C. 802) and § 555.11);

(5) Has been adjudicated as a mental defective or has been committed to a mental institution;

(6) Is an alien, other than an alien who—

(i) Is lawfully admitted for permanent residence (as that term is defined in section 101(a)(20) of the Immigration and Nationality Act (8 U.S.C. 1101)); or

(ii) Is in lawful nonimmigrant status, is a refugee admitted under section 207 of the Immigration and Nationality Act (8 U.S.C. 1157), or is in asylum status under section 208 of the Immigration and Nationality Act (8 U.S.C. 1158), and—

(A) Is a foreign law enforcement officer of a friendly foreign government, as determined by the Attorney General in consultation with the Secretary of State, entering the United States on official law enforcement business, and the shipping, transporting, possession, or receipt of explosive materials is in furtherance of this official law enforcement business;

(B) Is a person having the power to direct or cause the direction of the management and policies of a corporation, partnership, or association licensed pursuant to section 843(a) of the Act, and the shipping, transporting, possession, or receipt of explosive materials is in furtherance of such power;

(C) Is a member of a North Atlantic Treaty Organization (NATO) or other friendly foreign military force, as determined by the Attorney General in consultation with the Secretary of Defense, (whether or not admitted in a nonimmigrant status) who is present in the United States under military orders for training or other military purpose authorized by the United States, and the shipping, transporting, possession, or receipt of explosive materials is in furtherance of the military purpose; or

(D) Is lawfully present in the United States in cooperation with the Director of Central Intelligence, and the shipment, transportation, receipt, or possession of the explosive materials is in furtherance of such cooperation;

(7) Has been discharged from the armed forces under dishonorable conditions; or

(8) Having been a citizen of the United States, has renounced citizenship.

(e) See § 555.180 for regulations concerning the prohibited manufacture, importation, exportation, shipment, transportation, receipt, transfer, or possession of plastic explosives that do not contain a detection agent.

[ATF No. 1, 68 FR 13781, Mar. 20, 2003]

§ 555.27 Out-of-State disposition of explosive materials.

(a) No nonlicensee or nonpermittee may distribute any explosive materials to any other nonlicensee or nonpermittee who the distributor knows or who has reasonable cause to believe does not reside in the State in which the distributor resides.

(b) The provisions of this section do not apply on and after May 24, 2003.

[ATF No. 1, 68 FR 13782, Mar. 20, 2003]

§ 555.28 Stolen explosive materials.

No person shall receive, conceal, transport, ship, store, barter, sell, or dispose of any stolen explosive materials knowing or having reasonable cause to believe that the explosive materials were stolen.

§ 555.29 Unlawful storage.

No person shall store any explosive materials in a manner not in conformity with this part.

§ 555.30 Reporting theft or loss of explosive materials.

(a) Any licensee or permittee who has knowledge of the theft or loss of any explosive materials from his stock shall, within 24 hours of discovery, report the theft or loss by telephoning 1-800-800-3855 (nationwide toll free number) and on ATF F 5400.5 (formerly Form 4712) in accordance with the instructions on the form. Theft or loss of any explosive materials shall also be reported to appropriate local authorities.

(b) Any other person, except a carrier of explosive materials, who has knowledge of the theft or loss of any explosive materials from his stock shall, within 24 hours of discovery, report the theft or loss by telephoning 1-800-800-3855 (nationwide toll free number) and in writing to the nearest ATF office. Theft or loss shall be reported to appropriate local authorities.

(c) Reports of theft or loss of explosive materials under paragraphs (a) and (b) of this section must include the following information, if known:

(1) The manufacturer or brand name.

(2) The manufacturer's marks of identification (date and shift code).

(3) Quantity (applicable quantity units, such as pounds of explosives, number of detonators, etc.).

(4) Description (dynamite, blasting agents, detonators, etc.) and United Nations (UN) identification number, hazard division number, and classification letter, e.g., 1.1D, as classified by the U.S. Department of Transportation at 49 CFR 172.101 and 173.52.

(5) Size (length and diameter).

(d) A carrier of explosive materials who has knowledge of the theft or loss of any explosive materials shall, within 24 hours of discovery, report the theft or loss by telephoning 1-800-800-3855 (nationwide toll free number). Theft or loss shall also be reported to appropriate local authorities. Reports of theft or loss of explosive materials by carriers shall include the following information, if known:

(1) The manufacturer or brand name.

(2) Quantity (applicable quantity units, such as pounds of explosives, number of detonators, etc.).

(3) Description (United Nations (UN) identification number, hazard division number, and classification letter, e.g., 1.1D) as classified by the U.S. Department of Transportation at 49 CFR 172.101 and 173.52.

[T.D. ATF–87, 46 FR 40384, Aug. 7, 1981, as amended by T.D. ATF–400, 63 FR 45002, Aug. 24, 1998]

§ 555.31 Inspection of site accidents or fires; right of entry.

Any ATF officer may inspect the site of any accident or fire in which there is reason to believe that explosive materials were involved. Any ATF officer may enter into or upon any property where explosive materials have been used, are suspected of having been used, or have been found in an otherwise unauthorized location.

§ 555.32 Special explosive devices.

The Director may exempt certain explosive actuated devices, explosive actuated tools, or similar devices from the requirements of this part. A person who desires to obtain an exemption under this section for any special explosive device, which as designed does not constitute a public safety or security hazard, shall submit a written request to the Director. Each request shall be executed under the penalties of perjury and contain a complete and accurate description of the device, the name and address of the manufacturer or importer, the purpose of and use for which it is intended, and any photographs, diagrams, or drawings as may be necessary to enable the Director to make a determination. The Director may require that a sample of the device be submitted for examination and evaluation. If it is not possible to submit the device, the person requesting the exemption shall advise the Director and designate the place where the device will be available for examination and evaluation.

§ 555.33 Background checks and clearances (effective May 24, 2003).

(a) Background checks.

(1) If the Director receives from a licensee or permittee the names and appropriate identifying information of responsible persons and employees who will be authorized by the employer to possess explosive materials in the course of employment with the employer, the Director will conduct a background check in accordance with this section.

(2) The Director will determine whether the responsible person or employee is one of the persons described in any paragraph of section 842(i) of the Act (see § 555.26). In making such determination, the Director may take into account a letter or document issued under paragraph (a)(3) of this section.

(3) (i) If the Director determines that the responsible person or the employee is not one of the persons described in any paragraph of section 842(i) of the Act (see § 555.26), the Director will notify the employer in writing or electronically of the determination and issue, to the responsible person or employee, as the case may be, a letter of clearance which confirms the determination.

(ii) If the Director determines that the responsible person or employee is one of the persons described in any paragraph of section 842(i) of the Act (see § 555.26), ATF will notify the employer in writing or electronically of the determination and issue to the responsible person or the employee, as the case may be, a document that confirms the determination; explains the grounds for the determination; provides information on how the disability may be relieved; and explains how the determination may be appealed. The employer will retain the notification as part of his permanent records in accordance with § 555.121. The employer will take immediate steps to remove the responsible person from his position directing the management or policies of the business or operations as they relate to explosive materials or, as the case may be, to remove the employee from a position requiring the possession of explosive materials. Also, if the employer has listed the employee as a person authorized to accept delivery of explosive materials, as specified in § 555.103 or § 555.105, the employer must remove the employee from such list and immediately, and in no event later than the second business day after such change, notify distributors of such change.

(b) Appeals and correction of erroneous system information—

(1) In general. A responsible person or employee may challenge the adverse determination set out in the letter of denial, in writing and within 45 days of issuance of the determination, by directing his or her challenge to the basis for the adverse determination, or to the accuracy of the record upon which the adverse determination is based, to the Director. The appeal request must include appropriate documentation or record(s) establishing the legal and/or factual basis for the challenge. Any record or document of a court or other government entity or official furnished in support of an appeal must be certified by the court or other government entity or official as a true copy. In the case of an employee, or responsible person who did not submit fingerprints, such appeal must be accompanied by two properly completed FBI Forms FD–258 (fingerprint card). The Director will advise the individual in writing of his decision and the reasons for the decision.

(2) *Employees.* The letter of denial, among other things, will advise an employee who elects to challenge an adverse determination to submit the fingerprint cards as described above. The employee also will be advised of the agency name and address that originated the record containing the information causing the adverse determination ("originating agency"). At that time, and where appropriate, an employee is encouraged to apply to the originating agency to challenge the accuracy of the record(s) upon which the denial is based. The originating agency may respond to the individual's application by addressing the individual's specific reasons for the challenge, and by indicating whether additional information or documents are required. If the record is corrected as a result of the application to the originating agency, the individual may so notify ATF which will, in turn, verify the record correction with the originating agency and take all necessary steps to contact the agency responsible for the record system and correct the record. The employee may provide to ATF additional and appropriate documentation or record(s) establishing the legal and/or factual basis for the challenge to ATF's decision to uphold the initial denial. If ATF does not receive such additional documentation or record(s) within 45 days of the date of the decision upholding the initial denial, ATF will close the appeal.

(3) *Responsible persons.* The letter of denial, among other things, will advise a responsible person of the agency name and address which originated the record containing the information causing the adverse determination ("originating agency"). A responsible person who elects to challenge the adverse determination, where appropriate, is encouraged to apply to the originating agency to challenge the accuracy of the record(s) upon which the denial is based. The originating agency may respond to the individual's application by addressing the individual's specific reasons for the challenge, and by indicating whether additional information or documents are required. If the record is corrected as a result of the application to the originating agency, the individual may so notify ATF which will, in turn, verify the record correction with the originating agency and take all necessary steps to contact the agency responsible for the record system and correct the record. A responsible person may provide additional documentation or records as specified for employees in paragraph (b)(2) of this section.

(Approved by the Office of Management and Budget under control number 1140–0081)

[ATF No. 1, 68 FR 13783, Mar. 20, 2003]

§ 555.34 Replacement of stolen or lost ATF Form 5400.30 (Intrastate Purchase of Explosives Coupon (IPEC)).

When any Form 5400.30 is stolen, lost, or destroyed, the person losing possession will, upon discovery of the theft, loss, or destruction, immediately, but in all cases before 24 hours have elapsed since discovery, report the matter to the Director by telephoning 1-888-ATF-BOMB (nationwide toll free number). The report will explain in detail the circumstances of the theft, loss, or destruction and will include all known facts that may serve to identify the document. Upon receipt of the report, the Director will make such investigation as appears appropriate and may issue a duplicate document upon such conditions as the circumstances warrant.

(Approved by the Office of Management and Budget under control number 1140–0077)

[ATF No. 1, 68 FR 13783, Mar. 20, 2003]

Subpart D—Licenses and Permits

§ 555.41 General.

(a) *Licenses and permits issued prior to May 24, 2003.*

(1) Each person intending to engage in business as an importer or manufacturer of, or a dealer in, explosive materials, including black powder, must, before commencing business, obtain the license required by this subpart for the business to be operated. Each person who intends to acquire for use explosive materials from a licensee in a State other than the State in which he resides, or from a foreign country, or who intends to transport explosive materials in interstate or foreign commerce, must obtain a permit under this subpart; except that it is not necessary to obtain a permit if the user intends to lawfully purchase:

(i) Explosive materials from a licensee in a State contiguous to the user's State of residence and the user's State of residence has enacted legislation, currently in force, specifically authorizing a resident of that State to purchase explosive materials in a contiguous State; or

(ii) Commercially manufactured black powder in quantities not to exceed 50 pounds, intended to be used solely for sporting, recreational, or cultural purposes in antique firearms or in antique devices.

(2) Each person intending to engage in business as an explosive materials importer, manufacturer, or dealer must file an application, with the required fee (see § 555.42), with ATF in accordance with the instructions on the form (see § 555.45). A license will, subject to law, entitle the licensee to transport, ship, and receive explosive materials in interstate or foreign commerce, and to engage in the business specified by the license, at the location described on the license. A separate license must be obtained for each business premises at which the applicant is to manufacture, import, or distribute explosive materials except under the following circumstances:

(i) A separate license will not be required for storage facilities operated by the licensee as an integral part of one business premises or to cover a location used by the licensee solely for maintaining the records required by this part.

(ii) A separate license will not be required of a licensed manufacturer with respect to his on-site manufacturing.

(iii) It will not be necessary for a licensed importer or a licensed manufacturer (for purposes of sale or distribution) to also obtain a dealer's license in order to engage in business on his licensed premises as a dealer in explosive materials.

(iv) A separate license will not be required of licensed manufacturers with respect to their on-site manufacture of theatrical flash powder.

(3) Except as provided in paragraph (a)(1) of this section, each person intending to acquire explosive materials from a licensee in a State other than a State in which he resides, or from a foreign country, or who intends to transport explosive materials in interstate or foreign commerce, must file an application, with the required fee (see § 555.43), with ATF in accordance with the instructions on the form (see § 555.45). A permit will, subject to law, entitle the permittee to acquire, transport, ship, and receive in interstate or foreign commerce explosive materials. Only one permit is required under this part.

(b) Licenses and permits issued on and after May 24, 2003—

(1) In general.

(i) Each person intending to engage in business as an importer or manufacturer of, or a dealer in, explosive materials, including black powder, must, before commencing business, obtain the license required by this subpart for the business to be operated.

(ii) Each person who intends to acquire for use explosive materials within the State in which he resides on no more than 6 separate occasions during the 12-month period in which the permit is valid must obtain a limited permit under this subpart. (See § 555.105(b) for definition of "6 separate occasions.")

(iii) Each person who intends to acquire for use explosive materials from a licensee or permittee in a State other than the State in which he resides, or from a foreign country, or who intends to transport explosive materials in interstate or foreign commerce, or who intends to acquire for use explosive materials within the State in which he resides on more than 6 separate occasions during a 12-month period, must obtain a user permit under this subpart.

(iv) It is not necessary to obtain a permit if the user intends only to lawfully purchase commercially manufactured black powder in quantities not to exceed 50 pounds, intended to be used solely for sporting, recreational, or cultural purposes in antique firearms or in antique devices.

(2) Importers, manufacturers, and dealers. Each person intending to engage in business as an explosive materials importer, manufacturer, or dealer must file an application, with

the required fee (see § 555.42), with ATF in accordance with the instructions on the form (see § 555.45). A license will, subject to law, entitle the licensee to transport, ship, and receive explosive materials in interstate or foreign commerce, and to engage in the business specified by the license, at the location described on the license. A separate license must be obtained for each business premises at which the applicant is to manufacture, import, or distribute explosive materials except under the following circumstances:

(i) A separate license will not be required for storage facilities operated by the licensee as an integral part of one business premises or to cover a location used by the licensee solely for maintaining the records required by this part.

(ii) A separate license will not be required of a licensed manufacturer with respect to his on-site manufacturing.

(iii) It will not be necessary for a licensed importer or a licensed manufacturer (for purposes of sale or distribution) to also obtain a dealer's license in order to engage in business on his licensed premises as a dealer in explosive materials. No licensee will be required to obtain a user permit to lawfully transport, ship, or receive explosive materials in interstate or foreign commerce.

(iv) A separate license will not be required of licensed manufacturers with respect to their on-site manufacture of theatrical flash powder.

(3) Users of explosive materials.

(i) A limited permit will, subject to law, entitle the holder of such permit to receive for his use explosive materials from a licensee or permittee in his state of residence on no more than 6 separate occasions during the 12-month period in which the permit is valid. A limited permit does not authorize the receipt or transportation of explosive materials in interstate or foreign commerce. Holders of limited permits who need to receive explosive materials on more than 6 separate occasions during a 12-month period must obtain a user permit in accordance with this subpart.

(ii) Each person intending to acquire explosive materials from a licensee in a State other than a State in which he resides, or from a foreign country, or who intends to transport explosive materials in interstate or foreign commerce, must file an application for a user permit, with the required fee (see § 555.43), with ATF in accordance with the instructions on the form (see § 555.45). A user permit will, subject to law, entitle the permittee to transport, ship, and receive in interstate or foreign commerce explosive materials. Only one user permit per person is required under this part, irrespective of the number of locations relating to explosive materials operated by the holder of the user permit.

(Approved by the Office of Management and Budget under control number 1140–0083)

[ATF No. 1, 68 FR 13783, Mar. 20, 2003, as amended by ATF 5F, 70 FR 30633, May 27, 2005]

§ 555.42 License fees.

(a) Each applicant shall pay a fee for obtaining a three year license, a separate fee being required for each business premises, as follows:

    (1) Manufacturer-$200.

    (2) Importer-$200.

    (3) Dealer-$200.

(b) Each applicant for a renewal of a license shall pay a fee for a three year license as follows:

    (1) Manufacturer-$100.

    (2) Importer-$100.

    (3) Dealer-$100.

[T.D. ATF–400, 63 FR 45002, Aug. 24, 1998]

§ 555.43 Permit fees.

(a) Each applicant must pay a fee for obtaining a permit as follows:

    (1) User-$100 for a three-year period.

    (2) User-limited (nonrenewable)-$75.

    (3) Limited-$25 for a one-year period.

(b) **(1)** Each applicant for renewal of a user permit must pay a fee of $50 for a three-year period.

    (2) Each applicant for renewal of a limited permit must pay a fee of $12 for a one-year period.

[ATF No. 1, 68 FR 13785, Mar. 20, 2003]

§ 555.44 License or permit fee not refundable.

No refund of any part of the amount paid as a license or permit fee will be made where the operations of the licensee or permittee are, for any reason, discontinued during the period of an issued license or permit. However, the license or permit fee submitted with an application for a license or permit will be refunded if that application is denied, withdrawn, or abandoned, or if a license is cancelled subsequent to having been issued through administrative error.

§ 555.45 Original license or permit.

(a) Licenses issued prior to May 24, 2003. Any person who intends to engage in business as an explosive materials importer, manufacturer, or dealer, or who has not timely submitted application for renewal of a previous license issued under this part, shall file with ATF an application for License, Explosives, ATF F 5400.13/5400.16 with ATF in accordance with the instructions on the form. The application must be executed under the penalties of perjury and the penalties imposed by 18 U.S.C. 844(a). The application is to be accompanied by the appropriate fee in the form of a money order or check made payable to the Bureau of Alcohol, Tobacco and Firearms. ATF F 5400.13/5400.16 may be obtained from any ATF office. The Chief, Federal Explosives Licensing Center, will not approve an application postmarked on

or after March 20, 2003, unless it is submitted with a Responsible Person Questionnaire, ATF Form 5400.28. Form 5400.28 must be completed in accordance with the instructions on the form.

(b) Permits issued prior to May 24, 2003. Any person, except as provided in § 555.41(a), who intends to acquire explosive materials from a licensee in a state other than the State in which that person resides, or from a foreign country, or who intends to transport explosive materials in interstate or foreign commerce, or who has not timely submitted application for renewal of a previous permit issued under this part, shall file an application for Permit, Explosives, ATF F 5400.13/5400.16 or Permit, User Limited Special Fireworks, ATF F 5400.21 with ATF in accordance with the instructions on the form. The application must be executed under the penalties of perjury and the penalties imposed by 18 U.S.C. 844(a). The application is to be accompanied by the appropriate fee in the form of a money order or check made payable to the Bureau of Alcohol, Tobacco and Firearms. ATF F 5400.13/5400.16 and ATF F 5400.21 may be obtained from any ATF office. The Chief, Federal Explosives Licensing Center, will not approve an application postmarked on or after March 20, 2003, unless it is submitted with a Responsible Person Questionnaire, ATF Form 5400.28. Form 5400.28 must be completed in accordance with the instructions on the form.

(c) Licenses and permits issued on and after May 24, 2003—

    (1) License. Any person who intends to engage in the business as an importer of, manufacturer of, or dealer in explosive materials, or who has not timely submitted an application for renewal of a previous license issued under this part, must file an application for License, Explosives, ATF F 5400.13/5400.16, with ATF in accordance with the instructions on the form. ATF Form 5400.13/5400.16 may be obtained by contacting any ATF office. The application must:

        (i) Be executed under the penalties of perjury and the penalties imposed by 18 U.S.C. 844(a);

        (ii) Include appropriate identifying information concerning each responsible person;

        (iii) Include a photograph and fingerprints for each responsible person;

        (iv) Include the names of and appropriate identifying information regarding all employees who will be authorized by the applicant to possess explosive materials by submitting ATF F 5400.28 for each employee; and

        (v) Include the appropriate fee in the form of money order or check made payable to the Bureau of Alcohol, Tobacco, Firearms and Explosives.

    (2) User permit and limited permit. Except as provided in § 555.41(b)(1)(iv), any person who intends to acquire explosive materials in the State in which that person resides or acquire explosive materials from a licensee or holder of a user permit in a State other than the State in which that person resides, or from a foreign country, or who intends to transport explosive materials in interstate or foreign commerce, or who has not timely submitted an application for renewal of a previous permit issued under this part, must file an application for Permit, Explosives, ATF F 5400.13/5400.16 or Permit, User Limited Display Fireworks, ATF F 5400.21 with ATF in accordance with the instructions on the form. ATF Form 5400.13/5400.16 and ATF Form 5400.21 may be obtained by contacting any ATF office. The application must:

    (i)   Be executed under the penalties of perjury and the penalties imposed by 18 U.S.C. 844(a);

    (ii)   Include a photograph, fingerprints, and appropriate identifying information for each responsible person;

    (iii)   Include the names of and appropriate identifying information regarding all employees who will be authorized by the applicant to possess explosive materials by submitting ATF F 5400.28 for each employee; and

    (iv)   Include the appropriate fee in the form of money order or check made payable to the Bureau of Alcohol, Tobacco, Firearms and Explosives.

    (3)   The Federal Explosives Licensing Center, will conduct background checks on responsible persons and employees authorized by the applicant to possess explosive materials in accordance with § 555.33. If it is determined that any responsible person or employee is described in any paragraph of section 842(i) of the Act, the applicant must submit an amended application indicating removal or reassignment of that person before the license or permit will be issued.

    (Approved by the Office of Management and Budget under control number 1140–0083)

(18 U.S.C. 847 (84 Stat. 959); 18 U.S.C. 926 (82 Stat. 1226))

[T.D. ATF–200, 50 FR 10497, Mar. 15, 1985, as amended by T.D. ATF–400, 63 FR 45002, Aug. 24, 1998; ATF No. 1, 68 FR 13785, Mar. 20, 2003]

§ 555.46 Renewal of license or permit.

(a)   If a licensee or permittee intends to continue the business or operation described on a license or permit issued under this part after the expiration date of the license or permit, he shall, unless otherwise notified in writing by the Chief, Federal Explosives Licensing Center, execute and file prior to the expiration of his license or permit an application for license renewal, ATF F 5400.14 (Part III), or an application for permit renewal, ATF F 5400.15 (Part III), accompanied by the required fee, with ATF in accordance with the instructions on the form. In the event the licensee or permittee does not timely file a renewal application, he shall file an original application as required by § 555.45, and obtain the required license or permit in order to continue business or operations.

(b)   A user-limited permit is not renewable and is valid for a single purchase transaction. Applications for all user-limited permits must be filed on ATF F 5400.13/5400.16 or ATF F 5400.21, as required by § 555.45.

(18 U.S.C. 847 (84 Stat. 959); 18 U.S.C. 926 (82 Stat. 1226))

[T.D. ATF–87, 46 FR 40384, Aug. 7, 1981, as amended by T.D. ATF–200, 50 FR 10497, Mar. 15, 1985; T.D. ATF–290, 54 FR 53054, Dec. 27, 1989; T.D. ATF–400, 63 FR 45002, Aug. 24, 1998]

§ 555.47 Insufficient fee.

If an application is filed with an insufficient fee, the application and fee submitted will be returned to the applicant.

(18 U.S.C. 847 (84 Stat. 959); 18 U.S.C. 926 (82 Stat. 1226))

[T.D. ATF–200, 50 FR 10498, Mar. 15, 1985]

§ 555.48 Abandoned application.

Upon receipt of an incomplete or improperly executed application, the applicant will be notified of the deficiency in the application. If the application is not corrected and returned within 30 days following the date of notification, the application will be considered as having been abandoned and the license or permit fee returned.

§ 555.49 Issuance of license or permit.

(a)   Issuance of license or permit prior to May 24, 2003.

    (1)   The Chief, Federal Explosives Licensing Center, will issue a license or permit if—

    (I)   A properly executed application for the license or permit is received; and

    (ii)   Through further inquiry or investigation, or otherwise, it is found that the applicant is entitled to the license or permit.

    (2)   The Chief, Federal Explosives Licensing Center, will approve a properly executed application for a license or permit, if:

    (i)   The applicant is 21 years of age or over;

    (ii)   The applicant (including, in the case of a corporation, partnership, or association, any individual possessing, directly or indirectly, the power to direct or cause the direction of the management and policies of the corporation, partnership, or association) is not a person to whom distribution of explosive materials is prohibited under the Act;

    (iii)   The applicant has not willfully violated any provisions of the Act or this part;

    (iv)   The applicant has not knowingly withheld information or has not made any false or fictitious statement intended or likely to deceive, in connection with his application;

    (v)   The applicant has in a State, premises from which he conducts business or operations subject to license or permit under the Act or from which he intends to conduct business or operations;

    (vi)   The applicant has storage for the class (as described in § 555.202) of explosive materials described on the application, unless he establishes to the satisfaction of the Chief, Federal Explosives Licensing Center, that the business or operations to be conducted will not require the storage of explosive materials;

(vii) The applicant has certified in writing that he is familiar with and understands all published State laws and local ordinances relating to explosive materials for the location in which he intends to do business; and

(viii) The applicant for a license has submitted the certificate required by section 21 of the Federal Water Pollution Control Act, as amended (33 U.S.C. 1341).

(3) The Chief, Federal Explosives Licensing Center, will approve or the Director, Industry Operations will deny any application for a license or permit within the 45-day period beginning on the date a properly executed application was received. However, when an applicant for license or permit renewal is a person who is, under the provisions of § 555.83 or § 555.142, conducting business or operations under a previously issued license or permit, action regarding the application will be held in abeyance pending the completion of the proceedings against the applicant's existing license or permit, or renewal application, or final action by the Director on an application for relief submitted under § 555.142, as the case may be.

(4) The license or permit and one copy will be forwarded to the applicant, except that in the case of a user-limited permit, the original only will be issued.

(5) Each license or permit will bear a serial number and this number may be assigned to the licensee or permittee to whom issued for as long as he maintains continuity of renewal in the same region.

(b) Issuance of license or permit on and after May 24, 2003.

(1) The Chief, Federal Explosives Licensing Center, will issue a license or permit if:

(i) A properly executed application for the license or permit is received; and

(ii) Through further inquiry or investigation, or otherwise, it is found that the applicant is entitled to the license or permit.

(2) The Chief, Federal Explosives Licensing Center, will approve a properly executed application for a license or permit, if:

(i) The applicant (or, if the applicant is a corporation, partnership, or association, each responsible person with respect to the applicant) is not a person described in any paragraph of section 842(i) of the Act;

(ii) The applicant has not willfully violated any provisions of the Act or this part;

(iii) The applicant has not knowingly withheld information or has not made any false or fictitious statement intended or likely to deceive, in connection with his application;

(iv) The applicant has in a State, premises from which he conducts business or operations subject to license or permit under the Act or from which he intends to conduct business or operations;

(v) The applicant has storage for the class (as described in § 555.202) of explosive materials described on the application;

(vi) The applicant has certified in writing that he is familiar with and understands all published State laws and local ordinances relating to explosive materials for the location in which he intends to do business;

(vii) The applicant for a license has submitted the certificate required by section 21 of the Federal Water Pollution Control Act, as amended (33 U.S.C. 1341);

(viii) None of the employees of the applicant who will be authorized by the applicant to possess explosive materials is a person described in any paragraph of section 842(i) of the Act; and

(ix) In the case of an applicant for a limited permit, the applicant has certified in writing that the applicant will not receive explosive materials on more than 6 separate occasions during the 12-month period for which the limited permit is valid.

(3) The Chief, Federal Explosives Licensing Center, will approve or the Director, Industry Operations will deny any application for a license or permit within the 90-day period beginning on the date a properly executed application was received. However, when an applicant for license or permit renewal is a person who is, under the provisions of § 555.83 or § 555.142, conducting business or operations under a previously issued license or permit, action regarding the application will be held in abeyance pending the completion of the proceedings against the applicant's existing license or permit, or renewal application, or final action by the Director on an application for relief submitted under § 555.142, as the case may be.

(4) The license or permit and one copy will be forwarded to the applicant, except that in the case of a user-limited permit, the original only will be issued.

(5) Each license or permit will bear a serial number and this number may be assigned to the licensee or permittee to whom issued for as long as he maintains continuity of renewal in the same region.

(Approved by the Office of Management and Budget under control number 1140–0082)

[ATF No. 1, 68 FR 13785, Mar. 20, 2003]

§ 555.50 Correction of error on license or permit.

(a) Upon receipt of a license or permit issued under this part, each licensee or permittee shall examine the license or permit to insure that the information on it is accurate. If the license or permit is incorrect, the licensee or permittee shall return the license or permit to the Chief, Federal Explosives Licensing Center, with a statement showing the nature of the error. The Chief, Federal Explosives Licensing Center, shall correct the error, if the error was made in his office, and return the license or permit. However, if the error resulted from information contained in the licensee's or permittee's application for the license or permit, the Chief, Federal Explosives Licensing Center, shall require the licensee or permittee to file an amended application setting forth the correct information and a statement explaining the error contained in the application. Upon receipt of the amended application and a satisfactory explanation of the error, the Chief, Federal Explosives Licensing Center, shall make the correction on the license or permit and return it to the licensee or permittee.

(b) When the Chief, Federal Explosives Licensing Center, finds through any means other than notice from the licensee or permittee that an incorrect license or permit has been issued, (1) the Chief, Federal Explosives Licensing Center, may require the holder of the incorrect license or permit to return the license or permit for correction, and (2) if the error resulted from information contained in the licensee's or permittee's application for the license or permit, the Chief, Federal Explosives Licensing Center, shall require the licensee or permittee to file an amended application setting forth the correct information, and a statement satisfactorily explaining the error contained in the application. The Chief, Federal Explosives Licensing Center, then shall make the correction on the license or permit and return it to the licensee or permittee.

[T.D. ATF–87, 46 FR 40384, Aug. 7, 1981, as amended by T.D. ATF–290, 54 FR 53054, Dec. 27, 1989]

§ 555.51 Duration of license or permit.

(a) Prior to May 24, 2003. An original license or permit is issued for a period of three years. A renewal license or permit is issued for a period of three years. However, a user-limited permit is valid only for a single purchase transaction.

(b) On and after May 24, 2003

(1) An original license or user permit is issued for a period of three years. A renewal license or user permit is also issued for a period of three years. However, a user-limited permit is valid only for a single purchase transaction.

(2) A limited permit is issued for a period of one year. A renewal limited permit is also issued for a period of one year.

[ATF No. 1, 68 FR 13786, Mar. 20, 2003]

§ 555.52 Limitations on license or permit.

(a) The license covers the business of explosive materials specified in the license at the licensee's business premises (see § 555.41(b)).

(b) The permit is valid with respect to the type of operations of explosive materials specified in the permit.

[T.D. ATF–87, 46 FR 40384, Aug. 7, 1981, as amended by T.D. ATF–387, 62 FR 8376, Feb. 25, 1997; ATF 5F, 70 FR 30633, May 27, 2005]

§ 555.53 License and permit not transferable.

Licenses and permits issued under this part are not transferable to another person. In the event of the lease, sale, or other transfer of the business or operations covered by the license or permit, the successor must obtain the license or permit required by this part before commencing business or operations. However, for rules on right of succession, see § 555.59.

§ 555.54 Change of address.

(a) During the term of a license or permit, a licensee or permittee may move his business or operations to a new address at which he intends to regularly carry on his business or operations, without procuring a new license or permit. However, in every case, the licensee or permittee shall—

(1) Give notification of the new location of the business or operations to the Chief, Federal Explosives Licensing Center at least 10 days before the move; and

(2) Submit the license or permit to the Chief, Federal Explosives Licensing Center. The Chief, Federal Explosives Licensing Center will issue an amended license or permit, which will contain the new address (and new license or permit number, if any).

(b) Licensees and permittees whose mailing address will change must notify the Chief, Federal Explosives Licensing Center, at least 10 days before the change.

(Paragraph (b) approved by the Office of Management and Budget under control number 1140–0080)

[T.D. ATF–87, 46 FR 40384, Aug. 7, 1981, as amended by T.D. ATF–290, 54 FR 53054, Dec. 27, 1989; ATF No. 1, 68 FR 13786, Mar. 20, 2003]

§ 555.56 Change in trade name.

A licensee or permittee continuing to conduct business or operations at the location shown on his license or permit is not required to obtain a new license or permit by reason of a mere change in trade name under which he conducts his business or operations. However, the licensee or permittee shall furnish his license or permit and any copies furnished with the license or permit for endorsement of the change to the Chief, Federal Explosives Licensing Center, within 30 days from the date the licensee or permittee begins his business or operations under the new trade name.

[T.D. ATF–87, 46 FR 40384, Aug. 7, 1981, as amended by T.D. ATF–290, 54 FR 53054, Dec. 27, 1989]

§ 555.57 Change of control, change in responsible persons, and change of employees.

(a) In the case of a corporation or association holding a license or permit under this part, if actual or legal control of the corporation or association changes, directly or indirectly, whether by reason of change in stock ownership or control (in the corporation holding a license or permit or in any other corporation), by operation of law, or in any other manner, the licensee or permittee shall, within 30 days of the change, give written notification executed under the penalties of perjury, to the Chief, Federal Explosives Licensing Center. Upon expiration of the license or permit, the corporation or association shall file an ATF F 5400.13/5400.16 as required by § 555.45, and pay the fee prescribed in § 555.42(b) or § 555.43(b).

(b) For all licenses or permits issued on and after May 24, 2003, each person holding the license or permit must report to the Chief, Federal Explosives Licensing Center, any change in responsible persons or employees authorized to possess explosive materials. Such report must be submitted within 30 days of the change and must include appropriate identifying information for each responsible person. Reports relating to newly hired employees authorized to possess explosive materials must be submitted on ATF F 5400.28 for each employee.

(c) Upon receipt of a report, the Chief, Federal Explosives Licensing Center, will conduct a background check, if appropriate, in accordance with § 555.33.

(d) The reports required by paragraph (b) of this section must be retained as part of a licensee's or permittee's permanent records for the period specified in § 555.121.

(Approved by the Office of Management and Budget under control number 1140–0074)

[T.D. ATF–87, 46 FR 40384, Aug. 7, 1981, as amended by T.D. ATF–290, 54 FR 53054, Dec. 27, 1989; ATF No. 1, 68 FR 13786, Mar. 20, 2003]

§ 555.58 Continuing partnerships.

Where, under the laws of the particular State, the partnership is not terminated on death or insolvency of a partner, but continues until the winding up of the partnership affairs is completed, and the surviving partner has the exclusive right to the control and possession of the partnership assets for the purpose of liquidation and settlement, the surviving partner may continue to conduct the business or operations under the license or permit of the partnership. If the surviving partner acquires the business or operations on completion of settlement of the partnership, he shall obtain a license or permit in his own name from the date of acquisition, as provided in § 555.45. The rule set forth in this section will also apply where there is more than one surviving partner.

§ 555.59 Right of succession by certain persons.

(a) Certain persons other than the licensee or permittee may secure the right to carry on the same explosive materials business or operations at the same business premises for the remainder of the term of license or permit. These persons are:

(1) The surviving spouse or child, or executor, administrator, or other legal representative of a deceased licensee or permittee; and

(2) A receiver or trustee in bankruptcy, or an assignee for benefit of creditors.

(b) In order to secure the right of succession, the person or persons continuing the business or operations shall submit the license or permit and all copies furnished with the license or

28

permit for endorsement of the succession to the Chief, Federal Explosives Licensing Center, within 30 days from the date on which the successor begins to carry on the business or operations.

[T.D. ATF–87, 46 FR 40384, Aug. 7, 1981, as amended by T.D. ATF–290, 54 FR 53054, Dec. 27, 1989]

§ 555.60 Certain continuances of business or operations.

A licensee or permittee who furnishes his license or permit to the Chief, Federal Explosives Licensing Center, for correction, amendment, or endorsement, as provided in this subpart, may continue his business or operations while awaiting its return.

[T.D. ATF–87, 46 FR 40384, Aug. 7, 1981, as amended by T.D. ATF–290, 54 FR 53054, Dec. 27, 1989]

§ 555.61 Discontinuance of business or operations.

Where an explosive materials business or operations is either discontinued or succeeded by a new owner, the owner of the business or operations discontinued or succeeded shall, within 30 days, furnish notification of the discontinuance or succession and submit his license or permit and any copies furnished with the license or permit to the Federal Explosives Licensing Center. (See also § 555.128.)

[T.D. ATF–87, 46 FR 40384, Aug. 7, 1981, as amended by T.D. ATF–290, 54 FR 53054, Dec. 27, 1989]

§ 555.62 State or other law.

A license or permit issued under this part confers no right or privilege to conduct business or operations, including storage, contrary to State or other law. The holder of a license or permit issued under this part is not, by reason of the rights and privileges granted by that license or permit, immune from punishment for conducting an explosive materials business or operations in violation of the provisions of any State or other law. Similarly, compliance with the provisions of any State or other law affords no immunity under Federal law or regulations.

§ 555.63 Explosives magazine changes.

(a) General.

(1) The requirements of this section are applicable to magazines used for other than temporary (under 24 hours) storage of explosives.

(2) A magazine is considered suitable for the storage of explosives if the construction requirements of this part are met during the time explosives are stored in the magazine.

(3) A magazine is considered suitable for the storage of explosives if positioned in accordance with the applicable table of distances as specified in this part during the time explosives are stored in the magazine.

(4) For the purposes of this section, notification of the Director, Industry Operations may be by telephone or in writing. However, if notification of the Director, Industry Operations is in writing it must be at least three business days in advance of making changes in construction to an existing magazine or

constructing a new magazine, and at least five business days in advance of using any reconstructed magazine or added magazine for the storage of explosives.

(b) Exception. Mobile or portable type 5 magazines are exempt from the requirements of paragraphs (c) and (d) of this section, but must otherwise be in compliance with paragraphs (a)(2) and (3) of this section during the time explosives are stored in such magazines.

(c) Changes in magazine construction. A licensee or permittee who intends to make changes in construction of an existing magazine shall notify the Director, Industry Operations describing the proposed changes prior to making any changes. Unless otherwise advised by the Director, Industry Operations, changes in construction may commence after explosives are removed from the magazine. Explosives may not be stored in a reconstructed magazine before the Director, Industry Operations has been notified in accordance with paragraph (a)(4) of this section that the changes have been completed.

(d) Magazines acquired or constructed after permit or license is issued. A licensee or permittee who intends to construct or acquire additional magazines shall notify the Director, Industry Operations in accordance with paragraph (a)(4) of this section describing the additional magazines and the class and quantity of explosives to be stored in the magazine. Unless otherwise advised by the Director, Industry Operations, additional magazines may be constructed, or acquired magazines may be used for the storage of explosives. Explosives must not be stored in a magazine under construction. The Director, Industry Operations must be notified that construction has been completed.

[T.D. ATF–87, 46 FR 40384, Aug. 7, 1981, as amended by T.D. ATF–400, 63 FR 45002, Aug. 24, 1998]

Subpart E—License and Permit Proceedings

§ 555.71 Opportunity for compliance.

Except in cases of willfulness or those in which the public interest requires otherwise, and the Director, Industry Operations so alleges in the notice of denial of an application or revocation of a license or permit, no license or permit will be revoked or renewal application denied without first calling to the attention of the licensee or permittee the reasons for the contemplated action and affording him an opportunity to demonstrate or achieve compliance with all lawful requirements and to submit facts, arguments, or proposals of adjustment. The notice of contemplated action, ATF F 5400.12, will afford the licensee or permittee 15 days from the date of receipt of the notice to respond. If no response is received within the 15 days, or if after consideration of relevant matters presented by the licensee or permittee, the Director, Industry Operations finds that the licensee or permittee is not likely to abide by the law and regulations, he will proceed as provided in § 555.74.

[T.D. ATF–87, 46 FR 40384, Aug. 7 1981, as amended by T.D. ATF–446, 66 FR 16602, Mar. 27, 2001]

§ 555.72 Denial of initial application.

Whenever the Director, Industry Operations has reason to believe that an applicant for an original license or permit is not eligible to receive a license or permit under the provisions of § 555.49, he shall issue a notice of denial on ATF F 5400.11. The notice will set forth the matters of fact and law relied upon in determining that the application should be denied, and will afford the applicant 15 days from the date of receipt of the notice in which to request a hearing to review the denial. If no

request for a hearing is filed within that time, a copy of the application, marked *"Disapproved"*, will be returned to the applicant.

§ 555.73 Hearing after initial application is denied.

If the applicant for an original license or permit desires a hearing, he shall file a request with the Director, Industry Operations within 15 days after receipt of the notice of denial. The request should include a statement of the reasons for a hearing. On receipt of the request, the Director, Industry Operations shall refer the matter to an administrative law judge who shall set a time and place (see § 555.77) for a hearing and shall serve notice of the hearing upon the applicant and the Director, Industry Operations at least 10 days in advance of the hearing date. The hearing will be conducted in accordance with the hearing procedures prescribed in part 71 of this chapter (see § 555.82). Within a reasonable time after the conclusion of the hearing, and as expeditiously as possible, the administrative law judge shall render his recommended decision. He shall certify to the complete record of the proceedings before him and shall immediately forward the complete certified record, together with four copies of his recommended decision, to the Director, Industry Operations for decision.

§ 555.74 Denial of renewal application or revocation of license or permit.

If following the opportunity for compliance under § 555.71, or without opportunity for compliance under § 555.71, as circumstances warrant, the Director, Industry Operations finds that the licensee or permittee is not likely to comply with the law or regulations or is otherwise not eligible to continue operations authorized under his license or permit, the Director, Industry Operations shall issue a notice of denial of the renewal application or revocation of the license or permit, ATF F 5400.11 or ATF F 5400.10, as appropriate. In either case, the notice will set forth the matters of fact constituting the violations specified, dates, places, and the sections of law and regulations violated. The notice will, in the case of revocation of a license or permit, specify the date on which the action is effective, which date will be on or after the date the notice is served on the licensee or permittee. The notice will also advise the licensee or permittee that he may, within 15 days after receipt of the notice, request a hearing and, if applicable, a stay of the effective date of the revocation of his license or permit.

§555.75 Hearing after denial of renewal application or revocation of license or permit.

If a licensee or permittee whose renewal application has been denied or whose license or permit has been revoked desires a hearing, he shall file a request for a hearing with the Director, Industry Operations. In the case of the revocation of a license or permit, he may include a request for a stay of the effective date of the revocation. On receipt of the request the Director, Industry Operations shall advise the licensee or permittee whether the stay of the effective date of the revocation is granted. If the stay of the effective date of the revocation is granted, the Director, Industry Operations shall refer the matter to an administrative law judge who shall set a time and place (see § 555.77) for a hearing and shall serve notice of the hearing upon the licensee or permittee and the Director, Industry Operations at least 10 days in advance of the hearing date. If the stay of the effective date of the revocation is denied, the licensee or permittee may request an immediate hearing. In this event, the Director, Industry Operations shall immediately refer the matter to an administrative law judge who shall set a date and place for a hearing, which date shall be no later than 10 days from the date the licensee or permittee requested an immediate hearing. The hearing will be held in accordance with the applicable provisions of part 71 of this chapter. Within a reasonable time after the conclusion of the hearing, and as expeditiously as possible, the administrative law judge shall render his decision. He shall certify to the complete record of the proceeding before him and shall immediately forward the complete certified record, together with two copies of his decision, to the Director, Industry Operations, serve one copy of his decision on the licensee or permittee or his counsel, and transmit a copy to the attorney for the Government.

§555.76 Action by Director, Industry Operations.

(a) Initial application proceedings. If, upon receipt of the record and the recommended decision of the administrative law judge, the Director, Industry Operations decides that the license or permit should be issued, the Director, Industry Operations shall cause the application to be approved, briefly stating, for the record, his reasons. If he contemplates that the denial should stand, he shall serve a copy of the administrative law judge's recommended decision on the applicant, informing the applicant of his contemplated action and affording the applicant not more than 10 days in which to submit proposed findings and conclusions or exceptions to the recommended decision with supporting reasons. If the Director, Industry Operations, after consideration of the record of the hearing and of any proposed findings, conclusions, or exceptions filed with him by the applicant, approves the findings, conclusions and recommended decision of the administrative law judge, the Director, Industry Operations shall cause the license or permit to be issued or disapproved the application accordingly. If he disapproves the findings, conclusions, and recommendation of the administrative law judge, in whole or in part, he shall by order make such findings and conclusions as in his opinion are warranted by the law and the facts in the record. Any decision of the Director, Industry Operations ordering the disapproval of an initial application for a license or permit shall state the findings and conclusions upon which it is based, including his ruling upon each proposed finding, conclusion, and exception to the administrative law judge's recommended decision, together with a statement of his findings and conclusions, and reasons or basis for his findings and conclusions, upon all material issues of fact, law or discretion presented on the record. A signed duplicate original of the decision will be served upon the applicant and the original copy containing certificate of service will be placed in the official record of the proceedings. If the decision of the Director, Industry Operations is in favor of the applicant, he shall issue the license or permit, to be effective on issuance.

(b) Renewal application and revocation proceedings. Upon receipt of the complete certified records of the hearing, the Director, Industry Operations shall enter an order confirming the revocation of the license or permit, or disapproving the application, in accordance with the administrative law judge's findings and decision, unless he disagrees with the findings and decision. A signed duplicate original of the order, ATF F 5400.9, will be served upon the licensee or permittee and the original copy containing certificate of service will be placed in the official record of the proceedings. If the Director, Industry Operations disagrees with the findings and decision of the administrative law judge, he shall file a petition with the Director for review of the findings and decision, as provided in § 555.79. In either case, if the renewal application denial is sustained, a copy of the application marked *"Disapproved"* will be returned to the applicant. If the renewal application denial is reversed, a license or permit will be issued to become effective on expiration of the license or permit being renewed, or on the date of issuance, whichever is later. If the proceedings involve the revocation of a license or permit which expired before a decision is in favor of the licensee or permittee, the Director, Industry Operations shall:

(1) If renewal application was timely filed and a stay of the effective date of the revocation was granted, cause to be issued a license or permit effective on the date of issuance;

(2) If renewal application was not timely filed but a stay of the effective date of the revocation had been granted, request that a renewal application be filed and, following that, cause to be issued a license or permit to be effective on issuance; or

(3) If a stay of the effective date of the revocation had not been granted, request that an application be filed as provided in § 555.45, and process it in the same manner as for an application for an original license or permit.

(c) Curtailment of stay of revocation effective date. If, after approval of a request for a stay of the effective date of an order revoking a license or permit but before actions are completed under this subpart, the Director, Industry Operations finds that it is contrary to the public interest for the licensee or permittee to continue the operations or activities covered by his license or permit, the Director, Industry Operations may issue a notice of withdrawal of the approval, effective on the date of issuance. Notice of withdrawal will be served upon the licensee or permittee in the manner provided in § 555.81.

[T.D. ATF–87, 46 FR 40384, Aug. 7, 1981, as amended by T.D. ATF–290, 54 FR 53054, Dec. 27, 1989]

§ 555.77 Designated place of hearing.

The designated place of hearing set as provided in § 555.73 or § 555.75, will be at the location convenient to the aggrieved party.

§ 555.78 Representation at a hearing.

An applicant, licensee, or permittee may be represented by an attorney, certified public accountant, or other person recognized to practice before the Bureau of Alcohol, Tobacco, Firearms and Explosives as provided in 31 CFR Part 8, if he has otherwise complied with the applicable requirements of 26 CFR 601.521 through 601.527. The Director, Industry Operations shall be represented in proceedings under § § 555.73 and 555.75 by an attorney in the office of the chief counsel or regional counsel who is authorized to execute and file motions, briefs, and other papers in the proceedings, on behalf of the Director, Industry Operations, in his own name as *"Attorney for the Government"*.

[T.D. ATF–87, 46 FR 40384, Aug. 7, 1981, as amended by T.D. ATF–92, 46 FR 46916, Sept. 23, 1981]

§ 555.79 Appeal on petition to the Director.

An appeal to the Director is not required prior to filing an appeal with the U.S. Court of Appeals for judicial review. An appeal may be taken by the applicant, licensee, or permittee to the Director from a decision resulting from a hearing under § 555.73 or § 555.75. An appeal may also be taken by a Director, Industry Operations from a decision resulting from a hearing under § 555.75 as provided in § 555.76(b). The appeal shall be taken by filing a petition for review on appeal with the Director within 15 days of the service of an administrative law judge's decision or an order. The petition will set forth facts tending to show (a) action of an arbitrary nature, (b) action without reasonable warrant in fact, or (c) action contrary to law and regulations. A copy of the petition will be filed with the Director, Industry Operations or served on the applicant, licensee, or permittee, as the case may be. In the event of appeal, the Director, Industry Operations shall immediately forward the complete original record, by certified mail, to the Director for his consideration, review, and disposition as provided in subpart I of part 71 of this chapter. When, on appeal, the Director affirms the initial decision of the Director, Industry Operations or the administrative law judge, as the case may be, the initial decision will be final.

§ 555.80 Court review.

An applicant, licensee, or permittee may, within 60 days after receipt of the decision of the administrative law judge or the final order of the Director, Industry Operations or the Director, file a petition for a judicial review of the decision, with the U.S. Court of Appeals for the district in which he resides or has his principal place of business. The Director, upon notification that a petition has been filed, shall have prepared a complete transcript of the record of the proceedings. The Director, Industry Operations or the Director, as the case may be, shall certify to the correctness of the transcript of the record, forward one copy to the attorney for the Government in the review of the case, and file the original record of the proceedings with the original certificate in the U.S. Court of Appeals.

§ 555.81 Service on applicant, licensee, or permittee.

All notices and other formal documents required to be served on an applicant, licensee, or permittee under this subpart will be served by certified mail or by personal delivery. Where service is by personal delivery, the signed duplicate original copy of the formal document will be delivered to the applicant, licensee, or permittee, or, in the case of a corporation, partnership, or association, by delivering it to an officer, manager, or general agent, or to its attorney of record.

§ 555.82 Provisions of part 200 made applicable.

The provisions of subpart G of part 200 of this chapter, as well as those provisions of part 71 relative to failure to appear, withdrawal of an application or surrender of a permit, the conduct of hearings before an administrative law judge, and record of testimony, are hereby made applicable to application, license, and permit proceedings under this subpart to the extent that they are not contrary to or incompatible with this subpart.

§555.83 Operations by licensees or permittees after notice of denial or revocation.

In any case where a notice of revocation has been issued and a request for a stay of the effective date of the revocation has not been granted, the licensee or permittee shall not engage in the activities covered by the license or permit pending the outcome of proceedings under this subpart. In any case where notice of revocation has been issued but a stay of the effective date of the revocation has been granted, the licensee or permittee may continue to engage in the activities covered by his license or permit unless, or until, formally notified to the contrary: Provided, that in the event the license or permit would have expired before proceedings under this subpart are completed, timely renewal application must have been filed to continue the license or permit beyond its expiration date. In any case where a notice of denial of a renewal application has been issued, the licensee or permittee may continue to engage in the activities covered by the existing license or permit after the date of expiration of the license or permit until proceedings under this subpart are completed.

Subpart F—Conduct of Business or Operations

§555.101 Posting of license or user permit.

A license or user permit issued under this part, or a copy of a license or user permit, will be posted and available for inspection on the business premises at each place where explosive materials are manufactured, imported, or distributed.

[T.D. ATF–87, 46 FR 40384, Aug. 7, 1981. Redesignated by T.D. ATF–487, 68 FR 3748, Jan. 24, 2003. ATF No. 1, 68 FR 13786, Mar. 20, 2003]

§555.102 Authorized operations by permittees.

(a) In general. A permit issued under this part does not authorize the permittee to engage in the business of manufacturing, importing, or dealing in explosive materials. Accordingly, if a permittee's operations bring him within the definition of manufacturer, importer, or dealer under this part, he shall qualify for the appropriate license.

(b) Distributions of surplus stocks—

(1) Distributions of surplus stocks prior to May 24, 2003. Permittees are not authorized to engage in the business of sale or distribution of explosive materials. However, permittees may dispose of surplus stocks of explosive materials to other licensees or permittees in accordance with § 555.103, and to nonlicensees or to nonpermittees in accordance with § 555.105(a)(4).

(2) Distributions of surplus stocks on and after May 24, 2003. Permittees are not authorized to engage in the business of sale or distribution of explosive materials. However, permittees may dispose of surplus stocks of explosive materials to other licensees or permittees in accordance with § 555.103 and § 555.105.

[T.D. ATF–400, 63 FR 45002, Aug. 24, 1998, as amended by ATF No. 1, 68 FR 13787, Mar. 20, 2003]

§555.103 Transactions among licensees/permittees and transactions among licensees and holders of user permits.

(a) Transactions among licensees/permittees prior to May 24, 2003—

(1) **General.**

(i)   A licensed importer, licensed manufacturer or licensed dealer selling or otherwise distributing explosive materials (or a permittee disposing of surplus stock to a licensee or another permittee) who has the certified information required by this section may sell or distribute explosive materials to a licensee or permittee for not more than 45 days following the expiration date of the distributee's license or permit, unless the distributor knows or has reason to believe that the distributee's authority to continue business or operations under this part has been terminated.

(ii)   A licensed importer, licensed manufacturer or licensed dealer selling or otherwise distributing explosive materials (or a permittee disposing of surplus stock to another licensee or permittee) must verify the license or permit status of the distributee prior to the release of explosive materials ordered, as required by this section.

(iii) Licensees or permittees desiring to return explosive materials to a licensed manufacturer may do so without obtaining a certified copy of the manufacturer's license.

(iv)   Where possession of explosive materials is transferred at the distributor's premises, the distributor must in all instances verify the identity of the person accepting possession on behalf of the distributee before relinquishing possession. Before the delivery at the distributor's premises of explosive materials to an employee of a licensee or permittee, or to an employee of a common or contract carrier transporting explosive materials to a licensee or permittee, the distributor delivering explosive materials must obtain an executed ATF F 5400.8, Explosives Delivery Record, from the employee before releasing the explosive materials. The ATF F 5400.8 must contain all of the information required on the form and required by this part.

Example 1. An ATF F 5400.8 is required when:

a.   An employee of the purchaser takes possession at the distributor's premises.

b.   An employee of a common or contract carrier hired by the purchaser takes possession at the distributor's premises.

Example 2. An ATF F 5400.8 is not required when:

a.   An employee of the distributor takes possession of the explosives for the purpose of transport to the purchaser.

b.   An employee of a common or contract carrier hired by the distributor takes possession of the explosives for the purpose of transport to the purchaser.

(2) **License/permit verification of individuals.**

(i)   The distributee must furnish a certified copy (or, in the case of a user-limited, the original) of the license or permit. The certified copy need be furnished only once during the current term of the license or permit. Also, a licensee need not furnish certified copies of licenses to other licensed locations operated by such licensee.

(ii)   The distributor may obtain any additional verification as the distributor deems necessary.

(3) **License/permit verification of business organizations.**

(i)   A business organization may (in lieu of furnishing a certified copy of a license) furnish the distributor a certified list which contains the name, address, license number and date of license expiration of each licensed location. The certified list need be furnished only once during the current term of the license or permit. Also, a business organization need not furnish a certified list to other licensed locations operated by such business organization.

(ii)   A business organization must, prior to ordering explosive materials, furnish the licensee or permittee a current certified list of the representatives or agents authorized to order explosive materials on behalf of the business organization showing the name, address, and date and place of birth of each representative or agent. A licensee or permittee may not distribute explosive materials to a business organization on the order of a person who does not appear on the certified list of representatives or agents and, if the person does appear on the certified list, the licensee or permittee must verify the identity of such person.

(4) **Licensee/permittee certified statement.**

(i)   A licensee or permittee ordering explosive materials from another licensee or permittee must furnish a current, certified statement of the intended use of the explosive materials, e.g., resale, mining, quarrying, agriculture, construction, sport rocketry, road building, oil well drilling, seismographic research, to the distributor.

(ii)   For individuals, the certified statement of intended use must specify the name, address, date and place of birth, and social security number of the distributee.

(iii)   For business organizations, the certified statement of intended use must specify the taxpayer identification number, the identity and the principal and local places of business.

(iv)   The licensee or permittee purchasing explosive materials must revise the furnished copy of the certified statement only when the information is no longer current.

(5) **User-limited permit transactions.** A user-limited permit issued under the provisions of this part is valid for only a single purchase transaction and is not renewable (see § 555.51). Accordingly, at the time a user-limited permittee orders explosive materials, the licensed distributor must write on the front of the user-limited permit the transaction date, his signature, and the distributor's license number prior to returning the permit to the user-limited permittee.

(b)   Transactions among licensees/permittees on and after May 24, 2003—

(1) **General.**

(i)   A licensed importer, licensed manufacturer or licensed dealer selling or otherwise distributing explosive materials (or a holder of a user permit disposing of surplus stock to a licensee; a holder of a user permit; or a holder of a limited permit who is within the same State as the distributor) who has the certified information required by this section may sell or distribute explosive materials to a licensee or permittee for not more than 45 days following the expiration date of the distributee's license or permit, unless the distributor knows or has reason to believe that the distributee's authority to continue business or operations under this part has been terminated.

(ii)   A licensed importer, licensed manufacturer or licensed dealer selling or otherwise distributing explosive materials (or a holder of a user permit disposing of surplus stock to another licensee or permittee) must verify the license or permit status of the distributee prior to the release of explosive materials ordered, as required by this section.

(iii)   Licensees or permittees desiring to return explosive materials to a licensed manufacturer may do so without obtaining a certified copy of the manufacturer's license.

(2) **Verification of license/user permit.**

(i)   Prior to or with the first order of explosive materials, the distributee must provide the distributor a certified copy (or, in the case of a user-limited, the original) of the distributee's license or user permit. However, licensees or holders of user permits that are business organizations may (in lieu of a certified copy of a license or user permit) provide the distributor with a certified list that contains the name, address, license or user permit number, and date of the license or user permit expiration of each location.

(ii)   The distributee must also provide the distributor with a current list of the names of persons authorized to accept delivery of explosive materials on behalf of the distributee. The distributee ordering explosive materials must keep the list current and provide updated lists to licensees and holders of user permits on a timely basis. A distributor may not transfer possession of explosive materials to any person whose name does not appear on the current list of names of persons authorized to accept delivery of explosive materials on behalf of the distributee. Except as provided in paragraph (b)(3) of this section, in all instances the distributor must verify the identity of the distributee, or the employee of the distributee accepting possession of explosive materials on behalf of the distributee, by examining an identification document (as defined in § 555.11) before relinquishing possession.

    (iii) A licensee or holder of a user permit ordering explosive materials from another licensee or permittee must provide to the distributor a current, certified statement of the intended use of the explosive materials, e.g., resale, mining, quarrying, agriculture, construction, sport rocketry, road building, oil well drilling, seismographic research, etc.

    (A) For individuals, the certified statement of intended use must specify the name, address, date and place of birth, and social security number of the distributee.

    (B) For business organizations, the certified statement of intended use must specify the taxpayer identification number, the identity and the principal and local places of business.

    (C) The licensee or holder of a user permit purchasing explosive materials must revise the furnished copy of the certified statement only when the information is no longer current.

  (3) Delivery of explosive materials by common or contract carrier. When a common or contract carrier will transport explosive materials from a distributor to a distributee who is a licensee or holder of a user permit, the distributor must take the following actions before relinquishing possession of the explosive materials:

    (i) Verify the identity of the person accepting possession for the common or contract carrier by examining such person's valid, unexpired driver's license issued by any State, Canada, or Mexico; and

    (ii) Record the name of the common or contract carrier (i.e., the name of the driver's employer) and the full name of the driver. This information must be maintained in the distributor's permanent records in accordance with § 555.121.

  (4) User-limited permit transactions. A user-limited permit issued under the provisions of this part is valid for only a single purchase transaction and is not renewable (see § 555.51). Accordingly, at the time a user-limited permittee orders explosive materials, the licensed distributor must write on the front of the user-limited permit the transaction date, his signature, and the distributor's license number prior to returning the permit to the user-limited permittee.

(Approved by the Office of Management and Budget under control number 1140–0079)

[ATF No. 1, 68 FR 13787, Mar. 20, 2003, as amended by ATF No. 2, 68 FR 53512, Sept. 11, 2003]

§ 555.104 Certified copy of license or permit.

Except as provided in § 555.49(a), each person issued a license or permit under this part shall be furnished together with his license or permit a copy for his certification. If a person desires an additional copy of his license or permit for certification and for use under § 555.103, he shall:

(a) Make a reproduction of the copy of his license or permit and execute the certification on it;

(b) Make a reproduction of his license or permit, enter on the reproduction the statement: "I certify that this is a true copy of a (insert the word license or permit) issued to me to engage in the specified business or operations", and sign his name next to the statement; or

(c) Submit a request, in writing, for certified copies of his license or permit to the Chief, Federal Explosives Licensing Center. The request will show the name, trade name (if any), and address of the licensee or permittee and the number of copies of the license or permit desired. There is a fee of $1 for each copy of a license or permit issued by the Chief, Federal Explosives Licensing Center under this paragraph. Fee payment must accompany each request for additional copies of a license or permit. The fee must be paid by (1) cash, or (2) money order or check made payable to the Bureau of Alcohol, Tobacco, Firearms and Explosives.

[T.D. ATF–87, 46 FR 40384, Aug. 7, 1981, as amended by T.D. ATF–290, 54 FR 53054, Dec. 27, 1989]

§ 555.105 Distributions to nonlicensees, nonpermittees, and limited permittees.

(a) Distributions to nonlicensees and nonpermittees prior to May 24, 2003.

  (1) This section will apply in any case where distribution of explosive materials to the distributee is not otherwise prohibited by the Act or this part.

  (2) Except as provided in paragraph (a)(3) of this section, a licensed importer, licensed manufacturer, or licensed dealer may distribute explosive materials to a nonlicensee or nonpermittee if the nonlicensee or nonpermittee is a resident of the same State in which the licensee's business premises are located, and the nonlicensee or nonpermittee furnishes to the licensee the explosives transaction record, ATF F 5400.4, required by § 555.126. Disposition of ATF F 5400.4 will be made in accordance with § 555.126.

  (3) A licensed importer, licensed manufacturer, or licensed dealer may sell or distribute explosive materials to a resident of a State contiguous to the State in which the licensee's place of business is located if the purchaser's State of residence has enacted legislation, currently in force, specifically authorizing a resident of that State to purchase explosive materials in a contiguous State and the purchaser and the licensee have, prior to the distribution of the explosive materials, complied with all the requirements of paragraphs (a)(2), (a)(5), and (a)(6) of this section applicable to intrastate transactions occurring on the licensee's business premises.

(4) A permittee may dispose of surplus stocks of explosive materials to a nonlicensee or nonpermittee if the nonlicensee or nonpermittee is a resident of the same State in which the permittee's business premises or operations are located, or is a resident of a State contiguous to the State in which the permittee's place of business or operations are located, and if the requirements of paragraphs (a)(2), (a)(3), (a)(5), and (a)(6) of this section are fully met.

(5) A licensed importer, licensed manufacturer, or licensed dealer selling or otherwise distributing explosive materials to a business entity must verify the identity of the representative or agent of the business entity who is authorized to order explosive materials on behalf of the business entity. Each business entity ordering explosive materials must furnish the distributing licensee prior to or with the first order of explosive materials a current certified list of the names of representatives or agents authorized to order explosive materials on behalf of the business entity. The business entity ordering explosive materials is responsible for keeping the certified list current. A licensee may not distribute explosive materials to a business entity on the order of a person whose name does not appear on the certified list.

(6) Where the possession of explosive materials is transferred at the distributor's premises, the distributor must in all instances verify the identity of the person accepting possession on behalf of the distributee before relinquishing possession. Before the delivery at the distributor's premises of explosive materials to an employee of a nonlicensee or nonpermittee, or to an employee of a common or contract carrier transporting explosive materials to a nonlicensee or nonpermittee, the distributor delivering explosive materials must obtain an executed ATF F 5400.8 from the employee before releasing the explosive materials. The ATF F 5400.8 must contain all of the information required on the form and by this part. (See examples in § 555.103(a)).

(7) A licensee or permittee disposing of surplus stock may sell or distribute commercially manufactured black powder in quantities of 50 pounds or less to a nonlicensee or nonpermittee if the black powder is intended to be used solely for sporting, recreational, or cultural purposes in antique firearms as defined in 18 U.S.C. 921(a)(16), or in antique devices as exempted from the term "destructive device" in 18 U.S.C. 921(a)(4).

(b) Distributions to holders of limited permits on and after May 24, 2003.

(1) This section will apply in any case where distribution of explosive materials to the distributee is not otherwise prohibited by the Act or this part.

(2) A licensed importer, licensed manufacturer or a licensed dealer may distribute explosive materials to a holder of a limited permit if such permittee is a resident of the same State in which the licensee's business premises are located, the holder of the

limited permit presents in person or by mail ATF Form 5400.4, Limited Permittee Transaction Report (LPTR), and the licensee completes Form 5400.4 in accordance with § 555.126(b). In no event will a licensee distribute explosive materials to a holder of a limited permit unless the holder presents a Form 5400.4 with an original unaltered and unexpired Intrastate Purchase of Explosives Coupon (IPEC), ATF Form 5400.30, affixed. The coupon must bear the name, address, permit number, and the coupon number of the limited permittee seeking distribution of the explosives.

(3) A holder of a limited permit is authorized to receive explosive materials from a licensee or permittee whose premises are located in the same State of residence in which the premises of the holder of the limited permit are located on no more than 6 separate occasions during the one-year period of the permit. For purposes of this section, the term "6 separate occasions" means six deliveries of explosive materials. Each delivery must—

(i) Relate to a single purchase transaction made on one ATF F 5400.4;

(ii) Be referenced on one commercial invoice or purchase order; and

(iii) Be delivered to the holder of the limited permit in one shipment delivered at the same time.

(4) A holder of a user permit may dispose of surplus stocks of explosive materials to a licensee or holder of a user permit, or a holder of a limited permit who is a resident of the same State in which the premises of the holder of the user permit are located. A holder of a limited permit may dispose of surplus stocks of explosive materials to another holder of a limited permit who is a resident of the same State in which the premises of the distributor are located, if the transaction complies with the requirements of paragraph (b)(2) of this section and § 555.126(b). A holder of a limited permit may also dispose of surplus stocks of explosive materials to a licensee or holder of a user permit if the disposition occurs in the State of residence of the holder of the limited permit. (See § 555.103.)

(5) Each holder of a limited permit ordering explosive materials must furnish the distributing licensee prior to or with the first order of the explosive materials a current list of the names of employees authorized to accept delivery of explosive materials on behalf of the limited permittee. The distributee ordering explosive materials must keep the list current and provide updated lists to licensees and holders of user permits on a timely basis. A licensed importer, licensed manufacturer, licensed dealer, or permittee, selling or otherwise distributing explosive materials to a holder of a limited permit must, prior to delivering the explosive materials, obtain from the limited permittee a current list of persons who are authorized to accept deliveries of explosive materials on behalf of the limited permittee. A licensee or permittee may not deliver explosive materials to a person whose name does not appear on the list.

(6) (i) Delivery at the distributor's premises. Where possession of explosive materials is transferred directly to the distributee at the distributor's premises, the distributor must

obtain an executed Form 5400.4 in accordance with § 555.126(b) and must in all instances verify the identity of the person accepting possession on behalf of the distributee by examining an identification document (as defined in § 555.11) before relinquishing possession.

(ii) Delivery by distributor. Where possession of explosive materials is transferred by the distributor to the distributee away from the distributor's premises, the distributor must obtain an executed Form 5400.4 in accordance with § 555.126(b) and must in all instances verify the identity of the person accepting possession on behalf of the distributee by examining an identification document (as defined in § 555.11) before relinquishing possession.

(iii) Delivery by common or contract carrier hired by the distributor. Where a common or contract carrier hired by the distributor will transport explosive materials from the distributor to a holder of a limited permit:

(A) The limited permittee must, prior to delivery of the explosive materials, complete the appropriate section on Form 5400.4, affix to the Form 5400.4 one of the six IPECs he has been issued, and provide the form to the distributor in person or by mail.

(B) The distributor must, before relinquishing possession of the explosive materials to the common or contract carrier:

(1) Verify the identity of the person accepting possession for the common or contract carrier by examining such person's valid, unexpired driver's license issued by any State, Canada, or Mexico; and

(2) Record the name of the common or contract carrier (i.e., the name of the driver's employer) and the full name of the driver. This information must be maintained in the distributor's permanent records in accordance with § 555.121.

(C) At the time of delivery of the explosive materials, the common or contract carrier, as agent for the distributor, must verify the identity of the person accepting delivery on behalf of the distributee, note the type and number of the identification document (as defined in § 555.11) and provide this information to the distributor. The distributor must enter this information in the appropriate section on Form 5400.4.

(iv) Delivery by common or contract carrier hired by the distributee. Where a common or contract carrier hired by the distributee will transport explosive materials from the distributor to a holder of a limited permit:

(A) The limited permittee must, prior to delivery of the explosive materials, complete the appropriate section on Form 5400.4, affix to the Form 5400.4 one of the six IPECs he has been issued, and provide the form to the distributor in person or by mail.

(B) Before the delivery at the distributor's premises to the common or contract carrier who will transport explosive materials to the holder of a limited permit, the distributor must:

(1) Verify the identity of the person accepting possession for the common or contract carrier by examining such person's valid, unexpired driver's license issued by any State, Canada, or Mexico; and

(2) Record the name of the common or contract carrier (i.e., the name of the driver's employer) and the full name of the driver. This information must be maintained in the distributor's permanent records in accordance with § 555.121.

(7) A licensee or permittee disposing of surplus stock may sell or distribute commercially manufactured black powder in quantities of 50 pounds or less to a holder of a limited permit, nonlicensee, or nonpermittee if the black powder is intended to be used solely for sporting, recreational, or cultural purposes in antique firearms as defined in 18 U.S.C. 921(a)(16), or in antique devices as exempted from the term "destructive device" in 18 U.S.C. 921(a)(4).

(Approved by the Office of Management and Budget under control number 1140–0075)

[ATF No. 1, 68 FR 13788, Mar. 20, 2003, as amended by ATF No. 2, 68 FR 53513, Sept. 11, 2003]

§ 555.106 Certain prohibited distributions.

(a) A licensee or permittee may not distribute explosive materials to any person except—

(1) A licensee;

(2) A holder of a user permit; or

(3) A holder of a limited permit who is a resident of the State where distribution is made and in which the premises of the transferor are located.

(b) A licensee shall not distribute any explosive materials to any person:

(1) Who the licensee knows is less than 21 years of age;

(2) In any State where the purchase, possession, or use by a person of explosive materials would be in violation of any State law or any published ordinance applicable at the place of distribution;

(3) Who the licensee has reason to believe intends to transport the explosive materials into a State where the purchase, possession, or use of explosive materials is prohibited or which does not permit its residents to transport or ship explosive materials into the State or to receive explosive materials in the State; or

(4) Who the licensee has reasonable cause to believe intends to use the explosive materials for other than a lawful purpose.

(c) A licensee shall not distribute any explosive materials to any person knowing or having reason to believe that the person:

(1) Is, except as provided under § 555.142 (d) and (e), under indictment or information for, or was convicted in any court of, a crime punishable by imprisonment for a term exceeding 1 year;

(2) Is a fugitive from justice;

(3) Is an unlawful user of marijuana, or any depressant or stimulant drug, or narcotic drug (as these terms are defined in the Controlled Substances Act, 21 U.S.C. 802);

(4) Was adjudicated as a mental defective or was committed to a mental institution;

(5) Is an alien, other than an alien who—

(i) Is lawfully admitted for permanent residence (as that term is defined in section 101(a)(20) of the Immigration and Nationality Act (8 U.S.C. 1101));

(ii) Is in lawful nonimmigrant status, is a refugee admitted under section 207 of the Immigration and Nationality Act (8 U.S.C. 1157), or is in asylum status under section 208 of the Immigration and Nationality Act (8 U.S.C. 1158), and—

(A) Is a foreign law enforcement officer of a friendly foreign government, as determined by the Attorney General in consultation with the Secretary of State, entering the United States on official law enforcement business, and the shipping, transporting, possession, or receipt of explosive materials is in furtherance of this official law enforcement business;

(B) Is a person having the power to direct or cause the direction of the management and policies of a corporation, partnership, or association licensed pursuant to section 843(a), and the shipping, transporting, possession, or receipt of explosive materials is in furtherance of such power;

(C) Is a member of a North Atlantic Treaty Organization (NATO) or other friendly foreign military force, as determined by the Attorney General in consultation with the Secretary of Defense, (whether or not admitted in a nonimmigrant status) who is present in the United States under military orders for training or other military purpose authorized by the United States, and the shipping, transporting, possession, or receipt of explosive materials is in furtherance of the military purpose; or

(D) Is lawfully present in the United States in cooperation with the Director of Central Intelligence, and the shipment, transportation, receipt, or possession of the explosive materials is in furtherance of such cooperation;

(6) Has been discharged from the armed forces under dishonorable conditions; or

(7) Having been a citizen of the United States, has renounced citizenship.

(d) The provisions of this section do not apply to the purchase of commercially manufactured black powder in quantities not to exceed 50 pounds, intended to be used solely for sporting, recreational, or cultural purposes in antique firearms or in antique devices, if the requirements of § 555.105(a)(7) or (b)(7) are fully met.

[T.D. ATF–87, 46 FR 40384, Aug. 7, 1981. Redesignated by T.D. ATF–487, 68 FR 3748, Jan. 24, 2003. ATF No. 1, 68 FR 13790, Mar. 20, 2003]

§ 555.107 Record of transactions.

Each licensee and permittee shall keep records of explosive materials as required by subpart G of this part.

§ 555.108 Importation.

(a) Explosive materials imported or brought into the United States by a licensed importer or holder of a user permit may be released from customs custody to the licensed importer or holder of a user permit upon proof of his status as a licensed importer or holder of a user permit. Proof of status must be made by the licensed importer or holder of a user permit furnishing to the customs officer a certified copy of his license or permit (see § 555.103).

(b) A nonlicensee or nonpermittee may import or bring into the United States commercially manufactured black powder in quantities not to exceed 50 pounds. Upon submitting to the customs officer completed ATF F 5400.3, certifying that the black powder is intended to be used solely for sporting, recreational, or cultural purposes in antique firearms or in antique devices, black powder may be released from customs custody. The disposition of the executed ATF F 5400.3 will be in accordance with the instructions on the form.

(c) The provisions of this section are in addition to, and are not in lieu of, any applicable requirement under 27 CFR Part 447.

(d) For additional requirements relating to the importation of plastic explosives into the United States on or after April 24, 1997, see § 555.183.

(e) For requirements relating to the marking of imported explosive materials, see § 555.109.

[T.D. ATF–87, 46 FR 40384, Aug. 7, 1981, as amended by T.D. ATF–387, 62 FR 8376, Feb. 25, 1997; ATF No. 1, 68 FR 13790, Mar. 20, 2003; ATF 5F, 70 FR 30633, May 27, 2005]

§ 555.109 Identification of explosive materials.

(a) **General.** Explosive materials, whether manufactured in the United States or imported, must contain certain marks of identification.

(b) **Required marks—**

(1) Licensed manufacturers. Licensed manufacturers who manufacture explosive materials for sale or distribution must place the following marks of identification on explosive materials at the time of manufacture:

(i) The name of the manufacturer; and

(ii) The location, date, and shift of manufacture. Where a manufacturer operates his plant for only one shift during the day, he does not need to show the shift of manufacture.

(2) Licensed importers.

(i) Licensed importers who import explosive materials for sale or distribution must place the following marks of identification on the explosive materials they import:

(A) The name and address (city and state) of the importer; and

(B) The location (city and country) where the explosive materials were manufactured, date, and shift of manufacture. Where the foreign manufacturer operates his plant for only one shift during the day, he does not need to show the shift of manufacture.

(ii) Licensed importers must place the required marks on all explosive materials imported prior to distribution or shipment for use, and in no event later than 15 days after the date of release from Customs custody.

(c) General requirements.

(1) The required marks prescribed in this section must be permanent and legible.

(2) The required marks prescribed in this section must be in the English language, using Roman letters and Arabic numerals.

(3) Licensed manufacturers and licensed importers must place the required marks on each cartridge, bag, or other immediate container of explosive materials that they manufacture or import, as well as on any outside container used for the packaging of such explosive materials.

(4) Licensed manufacturers and licensed importers may use any method, or combination of methods, to affix the required marks to the immediate container of explosive materials, or outside containers used for the packaging thereof, provided the identifying marks are legible, permanent, show all the required information, and are not rendered unreadable by extended periods of storage.

(5) If licensed manufacturers or licensed importers desire to use a coding system and omit printed markings on the container that show all the required information specified in paragraphs (b)(1) and (2) of this section, they must file with ATF a letterhead application displaying the coding that they plan to use and explaining the manner of its application. The Director must approve the application before the proposed coding can be used.

(d) Exceptions—

(1) Blasting caps. Licensed manufacturers or licensed importers are only required to place the identification marks prescribed in this section on the containers used for the packaging of blasting caps.

(2) Alternate means of identification. The Director may authorize other means of identifying explosive materials, including fireworks, upon receipt of a letter application from the licensed manufacturer or licensed importer showing that such other identification is reasonable and will not hinder the effective administration of this part.

(Approved by the Office of Management and Budget under control numbers 1140–0055 and 1140–0062)

[ATF 5F, 70 FR 30633, May 27, 2005, as amended by ATF–11F, 73 FR 57242, Oct. 2, 2008]

§ 555.110 Furnishing of samples (Effective on and after January 24, 2003).

(a) In general. Licensed manufacturers and licensed importers and persons who manufacture or import explosive materials or ammonium nitrate must, when required by letter issued by the Director, furnish—

(1) Samples of such explosive materials or ammonium nitrate;

(2) Information on chemical composition of those products; and

(3) Any other information that the Director determines is relevant to the identification of the explosive materials or to identification of the ammonium nitrate.

(b) Reimbursement. The Director will reimburse the fair market value of samples furnished pursuant to paragraph (a) of this section, as well as reasonable costs of shipment.

(Approved by the Office of Management and Budget under control number 1140–0073)

[ATF No. 1, 68 FR 13790, Mar. 20, 2003]

Subpart G—Records and Reports

§ 555.121 General.

(a) (1) Licensees and permittees shall keep records pertaining to explosive materials in permanent form (i.e., commercial invoices, record books) and in the manner required in this subpart.

(2) Licensees and permittees shall keep records required by this part on the business premises for five years from the date a transaction occurs or until discontinuance of business or operations by the licensee or permittee. (See also § 555.128 for discontinuance of business or operations.)

(b) ATF officers may enter the premises of any licensee or holder of a user permit for the purpose of examining or inspecting any record or document required by or obtained under this part (see § 555.24). Section 843(f) of the Act requires licensees and holders of user permits to make all required records available for examination or inspection at all reasonable times. Section 843(f) of the Act also requires licensees and permittees (including holders of limited permits) to submit all reports and information relating to all required records and their contents, as the regulations in this part prescribe.

(c) Each licensee and permittee shall maintain all records of importation, production, shipment, receipt, sale, or other disposition, whether temporary or permanent, of explosive materials as the regulations in this part prescribe. Sections 842(f) and 842(g) of the Act make it unlawful for any licensee or permittee knowingly to make any false entry in, or fail to make entry in, any record required to be kept under the Act and the regulations in this part.

(Approved by the Office of Management and Budget under control number 1140–0030)

[T.D. ATF–87, 46 FR 40384, Aug. 7, 1981, as amended by T.D. ATF–172, 49 FR 14941, Apr. 16, 1984; ATF No. 1, 68 FR 13790, Mar. 20, 2003; ATF–11F, 73 FR 57242, Oct. 2, 2008]

§ 555.122 Records maintained by licensed importers.

(a) Each licensed importer shall take true and accurate physical inventories which will include all explosive materials on hand required to be accounted for in the records kept under this part. The licensed importer shall take a special inventory

(1) At the time of commencing business, which is the effective date of the license issued upon original qualification under this part;

(2) At the time of changing the location of his business to another region;

(3) At the time of discontinuing business; and

(4) At any time the Director, Industry Operations may in writing require. Each special inventory is to be prepared in duplicate, the original of which is submitted to the Director, Industry Operations, and the duplicate retained by the licensed importer. If a special inventory specified by paragraphs (a)(1) through (4) of this section has not been taken during the calendar year, at least one physical inventory will be taken. However, the record of the yearly inventory, other than a special inventory required by paragraphs (a)(1) through (4) of this section, will remain on file for inspection instead of being sent to the Director, Industry Operations. (See also § 555.127.)

(b) Each licensed importer shall, not later than the close of the next business day following the date of importation or other acquisition of explosive materials, enter the following information in a separate record:

(1) Date of importation or other acquisition.

(2) Name or brand name of manufacturer and country of manufacture.

(3) Manufacturer's marks of identification.

(4) Quantity (applicable quantity units, such as pounds of explosives, number of detonators, number of display fireworks, etc.).

(5) Description (dynamite (dyn), blasting agents (ba), detonators (det), display fireworks (df), etc.) and size (length and diameter or diameter only of display fireworks).

(c) Each licensed importer shall, not later than the close of the next business day following the date of distribution of any explosive materials to another licensee or a permittee, enter in a separate record the following information:

(1) Date of disposition.

(2) Name or brand name of manufacturer and country of manufacture.

(3) Manufacturer's marks of identification.

(4) Quantity (applicable quantity units, such as pounds of explosives, number of detonators, number of display fireworks, etc.).

(5) Description (dynamite (dyn), blasting agents (ba), detonators (det), display fireworks (df), etc.) and size (length and diameter or diameter only of display fireworks).

(6) License or permit number of licensee or permittee to whom the explosive materials are distributed.

(d) The Director, Industry Operations may authorize alternate records to be maintained by a licensed importer to record his distribution of explosive materials when it is shown by the licensed importer that alternate records will accurately and readily disclose the information required by paragraph (c) of this section. A licensed importer who proposes to use alternate records shall submit a letter application to the Director, Industry Operations and shall describe the proposed alternate records and the need for them. Alternate records are not to be employed by the licensed importer until approval is received from the Director, Industry Operations.

(e) Each licensed importer shall maintain separate records of the sales or other distribution made of explosive materials to nonlicensees or nonpermittees. These records are maintained as prescribed by § 555.126.

(Approved by the Office of Management and Budget under control number 1140–0030)

[T.D. ATF–87, 46 FR 40384, Aug. 7, 1981, as amended by T.D. ATF–172, 49 FR 14941, Apr. 16, 1984; T.D. ATF–293, 55 FR 3721, Feb. 5, 1990; T.D. ATF–400, 63 FR 45003, Aug. 24, 1998; ATF–11F, 73 FR 57242, Oct. 2, 2008]

§ 555.123 Records maintained by licensed manufacturers.

(a) Each licensed manufacturer shall take true and accurate physical inventories which will include all explosive materials on hand required to be accounted for in the records kept under this part. The licensed manufacturer shall take a special inventory

(1) At the time of commencing business, which is the effective date of the license issued upon original qualification under this part;

(2) At the time of changing the location of his premises to another region;

(3) At the time of discontinuing business; and

(4) At any other time the Director, Industry Operations may in writing require. Each special inventory is to be prepared in duplicate, the original of which is submitted to the Director, Industry Operations, and the duplicate retained by the licensed manufacturer. If a special inventory required by paragraphs (a)(1) through (4) of this section has not been taken during the calendar year, at least one physical inventory will be taken. However, the record of the yearly inventory, other than a special inventory

required by paragraphs (a) (1) through (4) of this section, will remain on file for inspection instead of being sent to the Director, Industry Operations. (See also § 555.127.)

(b) Each licensed manufacturer shall not later than the close of the next business day following the date of manufacture or other acquisition of explosive materials, enter the following information in a separate record:

(1) Date of manufacture or other acquisition.

(2) Manufacturer's marks of identification.

(3) Quantity (applicable quantity units, such as pounds of explosives, number of detonators, number of display fireworks, etc.).

(4) Name, brand name or description (dynamite (dyn), blasting agents (ba), detonators (det), display fireworks (df), etc.) and size (length and diameter or diameter only of display fireworks).

(c) Each licensed manufacturer shall, not later than the close of the next business day following the date of distribution of any explosive materials to another licensee or a permittee, enter in a separate record the following information:

(1) Date of disposition.

(2) Name or brand name of manufacturer or name of importer, as applicable, if acquired other than by his own manufacture.

(3) Manufacturer's marks of identification.

(4) Quantity (applicable quantity units, such as pounds of explosives, number of detonators, number of display fireworks, etc.).

(5) Description (dynamite (dyn), blasting agents (ba), detonators (det), display fireworks (df), etc.) and size (length and diameter or diameter only of display fireworks).

(6) License or permit number of licensee or permittee to whom the explosive materials are distributed.

(d) Each licensed manufacturer who manufactures explosive materials for his own use shall, not later than the close of the next business day following the date of use, enter in a separate record the following information:

(1) Date of use.

(2) Quantity (applicable quantity units, such as pounds of explosives, number of detonators, number of special fireworks, etc.).

(3) Description (dynamite (dyn), blasting agents (ba), detonators (det), display fireworks (df), etc.) and size (length and diameter or diameter only of display fireworks).

Exception: A licensed manufacturer is exempt from the recordkeeping requirements of this subsection if the explosive materials are manufactured for his own use and used within a 24 hour period at the same site.

(e) The Director, Industry Operations may authorize alternate records to be maintained by a licensed manufacturer to record his distribution or use of explosive materials when it is shown by the licensed manufacturer that alternate records will accurately and readily disclose the information required by paragraph (c) of this section. A licensed manufacturer who proposes to use alternate records shall submit a letter application to the Director, Industry Operations and shall describe the proposed alternate records and the need for them. Alternate records are not to be employed by the licensed manufacturer until approval is received from the Director, Industry Operations.

(f) Each licensed manufacturer shall maintain separate records of the sales or other distribution made of explosive materials to nonlicensees or nonpermittees. These records are maintained as prescribed by § 555.126.

(Approved by the Office of Management and Budget under control number 1140–0030)

[T.D. ATF–87, 46 FR 40384, Aug. 7, 1981, as amended by T.D. ATF–172, 49 FR 14941, Apr. 16, 1984; T.D. ATF–293, 55 FR 3721, Feb. 5, 1990; T.D. ATF–400, 63 FR 45003, Aug. 24, 1998; ATF–11F, 73 FR 57242, Oct. 2, 2008]

§ 555.124 Records maintained by licensed dealers.

(a) Each licensed dealer shall take true and accurate physical inventories which will include all explosive materials on hand required to be accounted for in the records kept under this part. The licensed dealer shall take a special inventory

(1) At the time of commencing business, which is the effective date of the license issued upon original qualification under this part;

(2) At the time of changing the location of his premises to another region;

(3) At the time of discontinuing business; and

(4) At any other time the Director, Industry Operations may in writing require. Each special inventory is to be prepared in duplicate, the original of which is submitted to the Director, Industry Operations, and the duplicate retained by the licensed dealer. If a special inventory required by paragraphs (a)(1) through (4) of this section has not been taken during the calendar year, at least one physical inventory will be taken. However, the record of the yearly inventory, other than a special inventory required by paragraphs (a)(1) through (4) of this section, will remain on file for inspection instead of being sent to the Director, Industry Operations. (See also § 555.127.)

(b) Each licensed dealer shall, not later than the close of the next business day following the date of purchase or other acquisition of explosive materials (except as provided in paragraph (d) of this section), enter the following information in a separate record:

(1) Date of acquisition.

(2) Name or brand name of manufacturer and name of importer (if any).

(3) Manufacturer's marks of identification.

(4) Quantity (applicable quantity units, such as pounds of explosives, number of detonators, number of display fireworks, etc.).

(5) Description (dynamite (dyn), blasting agents (ba), detonators (det), display fireworks (df), etc.) and size (length and diameter or diameter only of display fireworks).

(6) Name, address, and license or permit number of the person from whom the explosive materials are received.

(c) Each licensed dealer shall, not later than the close of the next business day following the date of use (if the explosives are used by the dealer) or the date of distribution of any explosive materials to another licensee or a permittee (except as provided in paragraph (d) of this section), enter in a separate record the following information:

(1) Date of disposition.

(2) Name or brand name of manufacturer and name of importer (if any).

(3) Manufacturer's marks of identification.

(4) Quantity (applicable quantity units, such as pounds of explosives, number of detonators, number of display fireworks, etc.).

(5) Description (dynamite (dyn), blasting agents (ba), detonators (det), display fireworks (df), etc.) and size (length and diameter or diameter only of display fireworks).

(6) License or permit number of licensee or permittee to whom the explosive materials are distributed.

(d) When a commercial record is kept by a licensed dealer showing the purchase or other acquisition information required for the permanent record prescribed by paragraph (b) of this section, or showing the distribution information required for the permanent record prescribed by paragraph (c) of this section, the licensed dealer acquiring or distributing the explosive materials may, for a period not exceeding seven days following the date of acquisition of distribution of the explosive materials, delay making the required entry into the permanent record of acquisition or distribution. However, until the required entry of acquisition or disposition is made in the permanent record, the commercial record must be (1) kept by the licensed dealer separate from other commercial documents kept by the licensee, and (2) readily available for inspection on the licensed premises.

(e) The Director, Industry Operations may authorize alternate records to be maintained by a licensed dealer to record his acquisition or disposition of explosive materials, when it is shown by the licensed dealer that alternate records will accurately and readily disclose the required information. A licensed dealer who proposes to use alternate records shall submit a letter application to the Director, Industry Operations and shall describe the proposed alternate records and the need for them. Alternate records are not to be employed by the licensed dealer until approval is received from the Director, Industry Operations.

(f) Each licensed dealer shall maintain separate records of the sales or other distribution made of explosive materials to nonlicensees or nonpermittees. These records are maintained as prescribed by § 555.126.

(Approved by the Office of Management and Budget under control number 1140–0030)

[T.D. ATF–87, 46 FR 40384, Aug. 7, 1981, as amended by T.D. ATF–172, 49 FR 14941, Apr. 16, 1984; T.D. ATF–293, 55 FR 3721, Feb. 5, 1990; T.D. ATF–400, 63 FR 45003, Aug. 24, 1998; ATF–11F, 73 FR 57242, Oct. 2, 2008]

§ 555.125 Records maintained by permittees.

(a) Records maintained by permittees prior to May 24, 2003.

(1) Each permittee must take true and accurate physical inventories that will include all explosive materials on hand required to be accounted for in the records kept under this part. The permittee must take a special inventory—

(i) At the time of commencing business, which is the effective date of the permit issued upon original qualification under this part;

(ii) At the time of changing the location of his premises to another region;

(iii) At the time of discontinuing business; and

(iv) At any other time the Director, Industry Operations may in writing require. Each special inventory is to be prepared in duplicate, the original of which is submitted to the Director, Industry Operations and the duplicate retained by the permittee. If a special inventory required by paragraphs (a)(1)(i) through (iv) of this section has not been taken during the calendar year, a permittee is required to take at least one physical inventory. However, the record of the yearly inventory, other than a special inventory required by paragraphs (a)(1)(i) through (iv) of this section, will remain on file for inspection instead of being sent to the Director, Industry Operations. (See also § 555.127.)

(2) Each permittee must, not later than the close of the next business day following the date of acquisition of explosive materials, enter the following information in a separate record:

(i) Date of acquisition;

(ii) Name or brand name of manufacturer;

(iii) Manufacturer's marks of identification;

(iv) Quantity (applicable quantity units, such as pounds of explosives, number of detonators, number of display fireworks, etc.);

(v) Description (dynamite (dyn), blasting agents (ba), detonators (det), display fireworks (df), etc., and size (length and diameter or diameter only of display fireworks)); and

(vi) Name, address, and license number of the persons from whom the explosive materials are received.

(3) Each permittee must, not later than the close of the next business day following the date of disposition of surplus explosive materials to another permittee or a licensee, enter in a separate record the information prescribed in § 555.124(c).

(4) Each permittee must maintain separate records of disposition of surplus stocks of explosive materials to nonlicensees or nonpermittees as prescribed in § 555.126.

(5) The Director, Industry Operations may authorize alternate records to be maintained by a permittee to record his acquisition of explosive materials, when it is shown by the permittee that alternate records will accurately and readily disclose the required information. A permittee who proposes to use alternate records must submit a letter application to the Director, Industry Operations and must describe the proposed alternate records and the need for them. Alternate records are not to be employed by the permittee until approval is received from the Director, Industry Operations.

(b) Records maintained by permittees on and after May 24, 2003.

(1) Each holder of a user permit must take true and accurate physical inventories that will include all explosive materials on hand required to be accounted for in the records kept under this part. The permittee must take a special inventory—

    (i) At the time of commencing business, which is the effective date of the permit issued upon original qualification under this part;

    (ii) At the time of changing the location of his premises;

    (iii) At the time of discontinuing business; and

    (iv) At any other time the Director, Industry Operations may in writing require. Each special inventory is to be prepared in duplicate, the original of which is submitted to the Director, Industry Operations and the duplicate retained by the permittee. If a special inventory required by paragraphs (b)(1)(i) through (iv) of this section has not been taken during the calendar year, a permittee is required to take at least one physical inventory. The record of the yearly inventory, other than a special inventory required by paragraphs (b)(1)(i) through (iv) of this section, will remain on file for inspection instead of being sent to the Director, Industry Operations. (See also § 555.127.)

(2) Each holder of a limited permit must take true and accurate physical inventories, at least annually, that will include all explosive materials on hand required to be accounted for in the records kept under this part.

(3) Each holder of a user permit or a limited permit must, not later than the close of the next business day following the date of acquisition of explosive materials, enter the following information in a separate record:

    (i) Date of acquisition;

    (ii) Name or brand name of manufacturer;

    (iii) Manufacturer's marks of identification;

    (iv) Quantity (applicable quantity units, such as pounds of explosives, number of detonators, number of display fireworks, etc.);

    (v) Description (dynamite (dyn), blasting agents (ba), detonators (det), display fireworks (df), etc., and size (length and diameter or diameter only of display fireworks)); and

    (vi) Name, address, and license number of the persons from whom the explosive materials are received.

(4) Each holder of a user permit or a limited permit must, not later than the close of the next business day following the date of disposition of surplus explosive materials to another permittee or a licensee, enter in a separate record the information prescribed in § 555.124(c).

(5) When a record book is used as a permittee's permanent record the permittee may delay entry of the required information for a period not to exceed seven days if the commercial record contains all of the required information prescribed by paragraphs (b)(3) and (b)(4) of this section. However, the commercial record may be used instead of a record book as a permanent record provided that the record contains all of the required information prescribed by paragraphs (b)(3) and (b)(4) of this section.

(6) Each holder of a user permit or a limited permit must maintain separate records of disposition of surplus stocks of explosive materials to holders of a limited permit as prescribed in § 555.126.

(7) The Director, Industry Operations may authorize alternate records to be maintained by a holder of a user permit or a limited permit to record his acquisition of explosive materials, when it is shown by the permittee that alternate records will accurately and readily disclose the required information. A permittee who proposes to use alternate records must submit a letter application to the Director, Industry Operations and must describe the proposed alternate records and the need for them. Alternate records are not to be employed by the permittee until approval is received from the Director, Industry Operations.

(Approved by the Office of Management and Budget under control number 1140–0030)

[ATF No. 1, 68 FR 13790, Mar. 20, 2003]

§ 555.126 Explosives transaction record for distribution of explosive materials prior to May 24, 2003 and Limited Permittee Transaction Report for distribution of explosive materials on and after May 24, 2003.

(a) Explosives transaction record for distribution of explosive materials prior to May 24, 2003.

(1) A licensee or permittee shall not temporarily or permanently distribute explosive materials to any person, other than another licensee or permittee, unless he records the transaction on an explosives transaction record, ATF F 5400.4.

(2) Before the distribution of explosive materials to a nonlicensee or nonpermittee who is a resident of the State in which the licensee or permittee maintains his business premises, or to a nonlicensee or nonpermittee who is not a resident of the State in which the licensee or permittee maintains his business premises and is acquiring explosive materials under § 555.105(a)(3), the licensee or permittee distributing the explosive materials shall obtain an executed ATF F 5400.4 from the distributee which contains all of the information required on the form and by the regulations in this part.

(3) Completed ATF F 5400.4 is to be retained by the licensee or permittee as part of his permanent records in accordance with paragraph (a)(4) of this section.

(4) Each ATF F 5400.4 is retained in numerical (by transaction serial number) order commencing with "1" and continuing in regular sequence. When the numbering of any series reaches "1,000,000," the licensee or permittee may recommence the series. The recommenced series is to be given an alphabetical prefix or suffix. Where there is a change in proprietorship, or in the individual, firm, corporate name or trade name, the series in use at the time of the change may be continued.

(5) The requirements of this section are in addition to any other recordkeeping requirement contained in this part.

(6) A licensee or permittee may obtain, upon request, a supply of ATF F 5400.4 from the Director.

(b) Limited Permittee Transaction Report for distribution of explosive materials on and after May 24, 2003.

(1) A licensee or permittee may not distribute explosive materials to any person who is not a licensee or permittee. A licensee or permittee may not distribute explosive materials to a limited permittee unless the distributor records the transaction on ATF Form 5400.4, Limited Permittee Transaction Report.

(2) Before distributing explosive materials to a limited permittee, the licensee or permittee must obtain an executed Form 5400.4 from the limited permittee with an original unaltered and unexpired Intrastate Purchase of Explosives Coupon (IPEC) affixed. Except when delivery of explosive materials is made by a common or contract carrier who is an agent of the limited permittee, the licensee, permittee, or an agent of the licensee or permittee, must verify the identity of the of the holder of the limited permit by examining an identification document (as defined in § 555.11) and noting on the Form 5400.4 the type of document presented. The licensee or permittee must complete the appropriate section on Form 5400.4 to indicate the type and quantity of explosive materials distributed, the license or permit number of the seller, and the date of the transaction. The licensee or permittee must sign and date the form and include any other information required by the instructions on the form and the regulations in this part.

(3) One copy of Form 5400.4 must be retained by the distributor as part of his permanent records in accordance with paragraph (b)(4) of this section and for the period specified in § 555.121. The distributor must mail the other copy of Form 5400.4 to the Bureau of Alcohol, Tobacco, Firearms and Explosives in accordance with the instructions on the form.

(4) Each Form 5400.4 must be retained in chronological order by date of disposition, or in alphabetical order by name of limited permittee. A licensee may not, however, use both methods in a single recordkeeping system. Where there is a change in proprietorship by a limited permittee, the forms may continue to be filed together after such change.

(5) The requirements of this section are in addition to any other recordkeeping requirement contained in this part.

(Approved by the Office of Management and Budget under control number 1140–0078)

[T.D. ATF–87, 46 FR 40384, Aug. 7, 1981, as amended by T.D. ATF–93, 46 FR 50787, Oct. 15, 1981; T.D. ATF–172, 49 FR 14941, Apr. 16, 1984; T.D. ATF–446, 66 FR 16602, Mar. 27, 2001; ATF No. 1, 68 FR 13791, Mar. 20, 2003]

§ 555.127 Daily summary of magazine transactions.

In taking the inventory required by §§§ 555.122, 555.123, 555.124, and 555.125, a licensee or permittee shall enter the inventory in a record of daily summary transactions to be kept at each magazine of an approved storage facility; however, these records may be kept at one central location on the business premises if separate records of daily transactions are kept for each magazine. Not later than the close of the next business day, each licensee and permittee shall record by manufacturer's name or brand name, the total quantity received in and removed from each magazine during the day, and the total remaining on hand at the end of the day. Quantity entries for display fireworks may be expressed as the number and size of individual display fireworks in a finished state or as the number of packaged display segments or packaged displays. Information as to the number and size of display fireworks contained in any one packaged display segment or packaged display shall be provided to any ATF officer on request. Any discrepancy which might indicate a theft or loss of explosive materials is to be reported in accordance with § 555.30.

[T.D. ATF–293, 55 FR 3722, Feb. 5, 1990, as amended by T.D. ATF–400, 63 FR 45003, Aug. 24, 1998]

§ 555.128 Discontinuance of business.

Where an explosive materials business or operations is discontinued and succeeded by a new licensee or new permittee, the records prescribed by this subpart shall appropriately reflect such facts and shall be delivered to the successor. Where discontinuance of the business or operations is absolute, the records required by this subpart must be delivered within 30 days following the business or operations discontinuance to any ATF office located in the region in which the business was located, or to the ATF Out-of-Business Records Center, 244 Needy Road, Martinsburg, West Virginia 25405. Where State law or local ordinance requires the delivery of records to other responsible authority, the Chief, Federal Explosives Licensing Center may arrange for the delivery of the records required by this subpart to such authority. (See also, § 555.61.)

[T.D. ATF–290, 54 FR 53054, Dec. 27, 1989, as amended by T.D. ATF–446a, 66 FR 19089, Apr. 13, 2001; ATF No. 1, 68 FR 13792, Mar. 20, 2003; ATF–11F, 73 FR 57242, Oct. 2, 2008]

§ 555.129 Exportation.

Exportation of explosive materials is to be in accordance with the applicable provisions of section 38 of the Arms Export Control Act (22 U.S.C. 2778) and implementing regulations. However, a licensed importer, licensed manufacturer, or licensed dealer exporting explosive materials shall maintain records showing the manufacture or acquisition of explosive materials as required by this part and records showing the quantity, the manufacturer's name or brand name of explosive materials, the name and address of the foreign consignee of the explosive materials, and the date the explosive materials were exported. See § 555.180 for regulations concerning the exportation of plastic explosives.

[T.D. ATF–87, 46 FR 40384, Aug. 7, 1981, as amended by T.D. ATF–387, 62 FR 8377, Feb. 25, 1997]

Subpart H—Exemptions

§ 555.141 Exemptions.

(a) **General.** Except for the provisions of §§ 555.180 and 555.181, this part does not apply to:

(1) Any aspect of the transportation of explosive materials via railroad, water, highway, or air which is regulated by the U.S. Department of Transportation and its agencies, and which pertains to safety. For example, regulations issued by the Department of Transportation addressing the security risk of aliens transporting explosives by commercial motor or railroad carrier from Canada preclude the enforcement of 18 U.S.C. 842(i)(5) against persons shipping, transporting, receiving, or possessing explosives incident to and in connection with the commercial transportation of explosives by truck or rail from Canada into the United States. Questions concerning this exception should be directed to ATF's Public Safety Branch [ATF's Explosives Industry Programs Branch] in Washington, DC.

(2) The use of explosive materials in medicines and medicinal agents in the forms prescribed by the official United States Pharmacopeia or the National Formulary. "The United States Pharmacopeia and The National Formulary," USP and NF Compendia, are available from the United States Pharmacopeial Convention, Inc., 12601 Twinbrook Parkway, Rockville, Maryland 20852.

(3) The transportation, shipment, receipt, or importation of explosive materials for delivery to any agency of the United States or to any State or its political subdivision.

(4) Small arms ammunition and components of small arms ammunition.

(5) The manufacture under the regulation of the military department of the United States of explosive materials for, or their distribution to or storage or possession by, the military or naval services or other agencies of the United States.

(6) Arsenals, navy yards, depots, or other establishments owned by, or operated by or on behalf of, the United States.

(7) The importation, distribution, and storage of fireworks classified as UN0336, UN0337, UN0431, or UN0432 explosives by the U.S. Department of Transportation at 49 CFR 172.101 and generally known as "consumer fireworks" or "articles pyrotechnic."

(8) Gasoline, fertilizers, propellant actuated devices, or propellant actuated industrial tools manufactured, imported, or distributed for their intended purposes.

(9) Industrial and laboratory chemicals which are intended for use as reagents and which are packaged and shipped pursuant to U.S. Department of Transportation regulations, 49 CFR Parts 100 to 177, which do not require explosives hazard warning labels.

(10) Model rocket motors that meet all of the following criteria—

(i) Consist of ammonium perchlorate composite propellant, black powder, or other similar low explosives;

(ii) Contain no more than 62.5 grams of total propellant weight; and

(iii) Are designed as single-use motors or as reload kits capable of reloading no more than 62.5 grams of propellant into a reusable motor casing.

(b) **Black powder.** Except for the provisions applicable to persons required to be licensed under subpart D, this part does not apply with respect to commercially manufactured black powder in quantities not to exceed 50 pounds, percussion caps, safety and pyrotechnic fuses, quills, quick and slow matches, and friction primers, if the black powder is intended to be used solely for sporting, recreational, or cultural purposes in antique firearms, as defined in 18 U.S.C. 921(a)(16) or antique devices, as exempted from the term "destructive devices" in 18 U.S.C. 921(a)(4).

[T.D. ATF–87, 46 FR 40384, Aug. 7, 1981 as amended by T.D. ATF–87, 46 FR 46916, Sept. 23, 1981; T.D. ATF–293, 55 FR 3722, Feb. 5, 1990; T.D. ATF–87, 62 FR 8377, Feb. 25, 1997; T.D. ATF–400, 63 FR 45003, Aug. 24, 1998; ATF No. 1, 68 FR 13792, Mar. 20, 2003; ATF 6F, 71 FR 46101, Aug. 11, 2006]

§ 555.142 Relief from disabilities (effective January 24, 2003).

(a) Any person prohibited from shipping or transporting any explosive in or affecting interstate or foreign commerce or from receiving or possessing any explosive which has been shipped or transported in or affecting interstate or foreign commerce may make application for relief from disabilities under section 845(b) of the Act.

(b) An application for relief from disabilities must be filed with the Director by submitting ATF Form 5400.29, Application for Restoration of Explosives Privileges, in accordance with the instructions on the form. The application must be supported by appropriate data, including the information specified in paragraph (f) of this section. Upon receipt of an incomplete or improperly executed application for relief, the applicant will be notified of the deficiency in the application. If the application is not corrected and returned within 30 days following the date of notification, the application will be considered abandoned.

(c) (1) The Director may grant relief to an applicant if it is established to the satisfaction of the Director that the circumstances regarding the disability and the applicant's record and reputation are such that the applicant will not be likely to act in a manner dangerous to public safety and that the granting of such relief is not contrary to the public interest.

(2) Except as provided in paragraph (c)(3) of this section, the Director will not grant relief if the applicant—

(i) Has not been discharged from parole or probation for a period of at least 2 years;

(ii) Is a fugitive from justice;

(iii) Is a prohibited alien;

(iv) Is an unlawful user of or addicted to any controlled substance;

(v) Has been adjudicated a mental defective or committed to a mental institution, unless the applicant was subsequently determined by a court, board, commission, or other lawful authority to have been restored to mental competency, to be no longer suffering from a mental disorder, and to have had all rights restored; or

(vi) Is prohibited by the law of the State where the applicant resides from receiving or possessing explosive materials.

(3) (i) The Director may grant relief to aliens who have been lawfully admitted to the United States or to persons who have not been discharged from parole or probation for a period of at least 2 years if he determines that the applicant has a compelling need to possess explosives, such as for purposes of employment.

(ii) The Director may grant relief to the persons identified in paragraph (c)(2) of this section in extraordinary circumstances where the granting of such relief is consistent with the public interest.

(d) A person who has been granted relief under this section is relieved of all disabilities imposed by the Act for the disabilities disclosed in the application. The granting of relief will not affect any disabilities incurred subsequent to the date the application was filed. Relief from disabilities granted to aliens will be effective only so long as the alien retains his or her lawful immigration status.

(e) (1) A licensee or permittee who is under indictment or information for, or convicted of, a crime punishable by imprisonment for a term exceeding one year during the term of a current license or permit, or while he has pending a license or permit renewal application, shall not be barred from licensed or permit operations for 30 days after the date of indictment or information or 30 days after the date upon which his conviction becomes final. Also, if he files his application for relief under this section within such 30 day period, he may further continue licensed or permit operations while his application is pending. A licensee or permittee who does not file an application within 30 days from the date of his indictment or information, or within 30 days from the date his conviction becomes final, shall not continue licensed or permit operations beyond 30 days from the date of his indictment or information or beyond 30 days from the date his conviction becomes final.

(2) In the event the term of a license or permit of a person expires during the 30 day period following the date of indictment of information of during the 30 day period after the date upon which his conviction becomes final or while his application for relief is pending, he shall file a timely application for renewal of his license or permit in order to continue licensed or permit operations. The license or permit application is to show that the applicant has been indicted or under information for, or convicted of, a crime punishable by imprisonment for a term exceeding one year.

(3) A licensee or permittee shall not continue licensed or permit operations beyond 30 days following the date the Director issues notification that the licensee's or permittee's application for removal of the disabilities resulting from an indictment, information or conviction has been denied.

(4) When a licensee or permittee may no longer continue licensed or permit operations under this section, any application for renewal of license of permit filed by the licensee or permittee while his application for removal of disabilities resulting from an indictment, information or conviction is pending, will be denied by the Director, Industry Operations.

(f) (1) Applications for relief from disabilities must include the following information:

(i) In the case of a corporation, or of any person having the power to direct or control the management of the corporation, information as to the absence of culpability in the offense for which the corporation, or any such person, was indicted, formally accused or convicted;

(ii) In the case of an applicant who is an individual, two properly completed FBI Forms FD 258 (fingerprint card), and a written statement from each of three references who are not related to the applicant by blood or marriage and have known the applicant for at least 3 years, recommending the granting of relief;

(iii) Written consent to examine and obtain copies of records and to receive statements and information regarding the applicant's background, including records, statements and other information concerning employment, medical history, military service, immigration status, and criminal record;

(iv) In the case of an applicant having been convicted of a crime punishable by imprisonment for a term exceeding one year, a copy of the indictment or information on which the applicant was convicted, the judgment of conviction or record of any plea of nolo contendere or plea of guilty or finding of guilt by the court;

(v) In the case of an applicant under indictment, a copy of the indictment or information;

(vi) In the case of an applicant who has been adjudicated a mental defective or committed to a mental institution, a copy of the order of a court, board, commission, or other lawful authority that made the adjudication or ordered the commitment, any petition that sought to have the applicant so adjudicated or committed, any medical records reflecting the reasons for commitment and diagnoses of the applicant, and any court order or finding of a court, board, commission, or other lawful authority showing the applicant's discharge from commitment, restoration of mental competency and the restoration of rights;

(vii) In the case of an applicant who has been discharged from the Armed Forces under dishonorable conditions, a copy of the applicant's Certificate of Release or Discharge from Active Duty (Department of Defense Form 214), Charge Sheet (Department of Defense Form 458), and final court martial order;

(viii) In the case of an applicant who, having been a citizen of the United States, has renounced his or her citizenship, a copy of the formal renunciation of nationality before a diplomatic or consular officer of the United States in a foreign state or before an officer designated by the Attorney General when the United States was in a state of war (see 8 U.S.C. 1481(a)(5) and (6)); and

(ix) In the case of an applicant who is an alien, documentation that the applicant is an alien who has been lawfully admitted to the United States; certification from the applicant including the applicant's INS-issued alien number or admission number, country/countries of citizenship, and immigration status, and certifying that the applicant is legally authorized to work in the United States, or other purposes for which possession of explosives is required; certification from an appropriate law enforcement agency of the applicant's country of citizenship stating that the applicant does not have a criminal record; and, if applicable, certification from a Federal explosives licensee or permittee or other employer stating that the applicant is employed by the employer and must possess explosive materials for purposes of employment. These certifications must be submitted in English.

(2) Any record or document of a court or other government entity or official required by paragraph (f)(1) of this section must be certified by the court or other government entity or official as a true copy.

(Approved by the Office of Management and Budget under control number 1140–0076)

[T.D. ATF–87, 46 FR 40384, Aug. 7, 1981. Redesignated by T.D. ATF–487, 68 FR 3748, Jan. 24, 2003. ATF No. 1, 68 FR 13792, Mar. 20, 2003]

Subpart I—Unlawful Acts, Penalties, Seizures and Forfeitures

§555.161 Engaging in business without a license.

Any person engaging in the business of importing, manufacturing, or dealing in explosive materials without a license issued under the Act, shall be fined not more than $10,000 or imprisoned not more than 10 years, or both.

§555.162 False statement or representation.

Any person who knowingly withholds information or makes any false or fictitious oral or written statement or furnishes or exhibits any false, fictitious, or misrepresented identification, intended or likely to deceive for the purpose of obtaining explosive materials, or a license, permit, exemption, or relief from disability under the Act, shall be fined not more than $10,000 or imprisoned not more than 10 years, or both.

§555.163 False entry in record.

Any licensed importer, licensed manufacturer, licensed dealer, or permittee who knowingly makes any false entry in any record required to be kept under subpart G of this part, shall be fined not more than $10,000 or imprisoned not more than 10 years, or both.

[T.D. ATF–87, 46 FR 40384, Aug. 7, 1981, as amended by T.D. ATF–400, 63 FR 45003, Aug. 24, 1998]

§555.164 Unlawful storage.

Any person who stores any explosive material in a manner not in conformity with this part, shall be fined not more than $1,000 or imprisoned not more than one year, or both.

§555.165 Failure to report theft or loss.

(a) Any person who has knowledge of the theft or loss of any explosive materials from his stock and fails to report the theft or loss within 24 hours of discovery in accordance with § 555.30, shall be fined not more than $1,000 or imprisoned not more than one year, or both.

(b) On and after January 24, 2003, any licensee or permittee who fails to report a theft of explosive materials in accordance with § 555.30 will be fined under title 18 U.S.C., imprisoned not more than 5 years, or both.

[T.D. ATF–87, 46 FR 40384, Aug. 7, 1981, as amended by ATF No. 1, 68 FR 13793, Mar. 20, 2003]

§555.166 Seizure or forfeiture.

Any explosive materials involved or used or intended to be used in any violation of the Act or of this part or in any violation of any criminal law of the United States are subject to seizure and forfeiture, and all provisions of title 26, U.S.C. relating to the seizure, forfeiture, and disposition of firearms, as defined in 26 U.S.C. 5845(a), will, so far as applicable, extend to seizures and forfeitures under the Act. (See § 72.27 of this title for regulations on summary destruction of explosive materials which are impracticable or unsafe to remove to a place of storage.)

[T.D. ATF–87, 46 FR 40384, Aug. 7, 1981, as amended by T.D. ATF–363, 60 FR 17449, Apr. 6, 1995]

Subpart J—Marking of Plastic Explosives

§555.180 Prohibitions relating to unmarked plastic explosives.

(a) No person shall manufacture any plastic explosive that does not contain a detection agent.

(b) No person shall import or bring into the United States, or export from the United States, any plastic explosive that does not contain a detection agent. This paragraph does not apply to the importation or bringing into the United States, or the exportation from the United States, of any plastic explosive that was imported or brought into, or manufactured in the United States prior to April 24, 1996, by or on behalf of any agency of the United States performing military or police functions (including any military reserve component) or by or on behalf of the National Guard of any State, not later than 15 years after the date of entry into force of the Convention on the Marking of Plastic Explosives with respect to the United States, i.e., not later than June 21, 2013.

(c) No person shall ship, transport, transfer, receive, or possess any plastic explosive that does not contain a detection agent. This paragraph does not apply to:

(1) The shipment, transportation, transfer, receipt, or possession of any plastic explosive that was imported or brought into, or manufactured in the United States prior to April 24, 1996, by any person during the period beginning on that date and ending on April 24, 1999; or

(2) The shipment, transportation, transfer, receipt, or possession of any plastic explosive that was imported or brought into, or manufactured in the United States prior to April 24, 1996, by or on behalf of any agency of the United States performing a military or police function (including any military reserve component) or by or on behalf of the National Guard of any State, not later than 15 years after the date of entry into force of the Convention on the Marking of Plastic Explosives with respect to the United States, i.e., not later than June 21, 2013.

(d) When used in this subpart, terms are defined as follows:

(1) **Convention on the Marking of Plastic Explosives** means the Convention on the Marking of Plastic Explosives for the Purposes of Detection, Done at Montreal on 1 March 1991.

(2) **"Date of entry into force" of the Convention on the Marking of Plastic Explosives** means that date on which the Convention enters into force with respect to the U.S. in accordance with the provisions of Article XIII of the Convention on the Marking of Plastic Explosives. The Convention entered into force on June 21, 1998.

(3) **Detection agent** means any one of the substances specified in this paragraph when introduced into a plastic explosive or formulated in such explosive as a part of the manufacturing process in such a manner as to achieve homogeneous distribution in the finished explosive, including—

(i) Ethylene glycol dinitrate (EGDN), $C_2H_4(NO_3)_2$, molecular weight 152, when the minimum concentration in the finished explosive is 0.2 percent by mass;

(ii) 2,3-Dimethyl-2,3-dinitrobutane (DMNB), $C_6H_{12}(NO_2)_2$, molecular weight 176, when the minimum concentration in the finished explosive is 0.1 percent by mass;

(iii) Para-Mononitrotoluene (p-MNT), $C_7H_7NO_2$, molecular weight 137, when the minimum concentration in the finished explosive is 0.5 percent by mass;

(iv) Ortho-Mononitrotoluene (o-MNT), $C_7H_7NO_2$, molecular weight 137, when the minimum concentration in the finished explosive is 0.5 percent by mass; and

(v) Any other substance in the concentration specified by the Director, after consultation with the Secretary of State and Secretary of Defense, that has been added to the table in Part 2 of the Technical Annex to the Convention on the Marking of Plastic Explosives.

(4) **Plastic explosive** means an explosive material in flexible or elastic sheet form formulated with one or more high explosives which in their pure form has a vapor pressure less than 10^4 Pa at a temperature of 25 °C, is formulated with a binder material, and is as a mixture malleable or flexible at normal room temperature. High explosives, as defined in § 555.202(a), are explosive materials which can be caused to detonate by means of a blasting cap when unconfined.

[T.D. ATF–387, 62 FR 8376, Feb. 25, 1997, as amended by T.D. ATF–419, 64 FR 55628, Oct. 14, 1999]

§ 555.181 Reporting of plastic explosives.

All persons, other than an agency of the United States (including any military reserve component) or the National Guard of any State, possessing any plastic explosive on April 24, 1996, shall submit a report to the Director no later than August 22, 1996. The report shall be in writing and mailed by certified mail (return receipt requested) to the Director at P.O. Box 50204, Washington, DC 20091-0204. The report shall include the quantity of plastic explosives possessed on April 24, 1996; any marks of identification on such explosives; the name and address of the manufacturer or importer; the storage location of such explosives, including the city and State; and the name and address of the person possessing the plastic explosives.

[T.D. ATF–382, 61 FR 38085, July 23, 1996, as amended by T.D. ATF–387, 62 FR 8377, Feb. 25, 1997; ATF–11F, 73 FR 57242, Oct. 2, 2008]

§ 555.182 Exceptions.

It is an affirmative defense against any proceeding involving §§ 555.180 and 555.181 if the proponent proves by a preponderance of the evidence that the plastic explosive—

(a) Consisted of a small amount of plastic explosive intended for and utilized solely in lawful—

    (1) Research, development, or testing of new or modified explosive materials;

    (2) Training in explosives detection or development or testing of explosives detection equipment; or

    (3) Forensic science purposes; or

(b) Was plastic explosive that, by April 24, 1999, will be or is incorporated in a military device within the territory of the United States and remains an integral part of such military device, or is intended to be, or is incorporated in, and remains an integral part of a military device that is intended to become, or has become, the property of any agency of the United States performing military or police functions (including any military reserve component) or the National Guard of any State, wherever such device is located. For purposes of this paragraph, the term "military device" includes, but is not restricted to, shells, bombs, projectiles, mines, missiles, rockets, shaped charges, grenades, perforators, and similar devices lawfully manufactured exclusively for military or police purposes.

[T.D. ATF–387, 62 FR 8377, Feb. 25, 1997]

§ 555.183 Importation of plastic explosives on or after April 24, 1997.

Persons filing Form 6 applications for the importation of plastic explosives on or after April 24, 1997, shall attach to the application the following written statement, prepared in triplicate, executed under the penalties of perjury:

(a) "I declare under the penalties of perjury that the plastic explosive to be imported contains a detection agent as required by 27 CFR 555.180(b)"; or

(b) "I declare under the penalties of perjury that the plastic explosive to be imported is a "small amount" to be used for research, training, or testing purposes and is exempt from the detection agent requirement pursuant to 27 CFR 555.182."

[T.D. ATF–387, 62 FR 8377, Feb. 25, 1997]

§ 555.184 Statements of process and samples.

(a) A complete and accurate statement of process with regard to any plastic explosive or to any detection agent that is to be introduced into a plastic explosive or formulated in such plastic explosive shall be submitted by a licensed manufacturer or licensed importer, upon request, to the Director.

(b) Samples of any plastic explosive or detection agent shall be submitted by a licensed manufacturer or licensed importer, upon request, to the Director.

(Paragraph (a) approved by the Office of Management and Budget under control number 1140–0042)

[T.D. ATF–387, 62 FR 8378, Feb. 25, 1997, as amended by ATF–11F, 73 FR 57242, Oct. 2, 2008]

§ 555.185 Criminal sanctions.

Any person who violates the provisions of 18 U.S.C. 842(l)–(o) shall be fined under title 18, U.S.C., imprisoned for not more than 10 years, or both.

[T.D. ATF–387, 62 FR 8378, Feb. 25, 1997]

§ 555.186 Seizure or forfeiture.

Any plastic explosive that does not contain a detection agent in violation of 18 U.S.C. 842(l)–(n) is subject to seizure and forfeiture, and all provisions of 19 U.S.C. 1595a, relating to seizure, forfeiture, and disposition of merchandise introduced or attempted to be introduced into the U.S. contrary to law, shall extend to seizures and forfeitures under this subpart. See § 72.27 of this chapter for regulations on summary destruction of plastic explosives that do not contain a detection agent.

[T.D. ATF–387, 62 FR 8378, Feb. 25, 1997]

Subpart K—Storage

§ 555.201 General.

(a) Section 842(j) of the Act and § 555.29 of this part require that the storage of explosive materials by any person must be in accordance with the regulations in this part. Further, section 846 of this Act authorizes regulations to prevent the recurrence of accidental explosions in which explosive materials were involved. The storage standards prescribed by this subpart confer no right or privileges to store explosive materials in a manner contrary to State or local law.

(b) The Director may authorize alternate construction for explosives storage magazines when it is shown that the alternate magazine construction is substantially equivalent to the standards of safety and security contained in this subpart. Any alternate explosive magazine construction approved by the Director prior to August 9, 1982, will continue as approved unless notified in writing by the Director. Any person intending to use alternate magazine construction shall submit a letter application to the Director, Industry Operations for transmittal to the Director, specifically describing the proposed magazine. Explosive materials may not be stored in alternate magazines before the applicant has been notified that the application has been approved.

(c) A licensee or permittee who intends to make changes in his magazines, or who intends to construct or acquire additional magazines, shall comply with § 555.63.

(d) The regulations set forth in §§ 555.221 through 555.224 pertain to the storage of display fireworks, pyrotechnic compositions, and explosive materials used in assembling fireworks and articles pyrotechnic.

(e) The provisions of § 555.202(a) classifying flash powder and bulk salutes as high explosives are mandatory after March 7, 1990: Provided, that those persons who hold licenses or permits under this part on that date shall, with respect to the premises covered by such licenses or permits, comply with the high explosives storage requirements for flash powder and bulk salutes by March 7, 1991.

(f) Any person who stores explosive materials shall notify the authority having jurisdiction for fire safety in the locality in which the explosive materials are being stored of the type, magazine capacity, and location of each site where such explosive materials are stored. Such notification shall be made orally before the end of the day on which storage of the explosive materials commenced and in writing within 48 hours from the time such storage commenced.

(Paragraph (f) approved by the Office of Management and Budget under control number 1140–0071)

[T.D. ATF–87, 46 FR 40384, Aug. 7, 1981, as amended by T.D. ATF–293, 55 FR 3722, Feb. 5, 1990; T.D. ATF–400, 63 FR 45003, Aug. 24, 1998; ATF–11F, 73 FR 57242, Oct. 2, 2008]

§ 555.202 Classes of explosive materials.

For purposes of this part, there are three classes of explosive materials. These classes, together with the description of explosive materials comprising each class, are as follows:

(a) **High explosives.** Explosive materials which can be caused to detonate by means of a blasting cap when unconfined, (for example, dynamite, flash powders, and bulk salutes). See also § 555.201(c).

(b) **Low explosives.** Explosive materials which can be caused to deflagrate when confined (for example, black powder, safety fuses, igniters, igniter cords, fuse lighters, and "display fireworks" classified as UN0333, UN0334, or UN0335 by the U.S. Department of Transportation regulations at 49 CFR 172.101, except for bulk salutes).

(c) **Blasting agents.** (For example, ammonium nitrate-fuel oil and certain water-gels (see also § 555.11).

[T.D. ATF–87, 46 FR 40384, Aug. 7, 1981, as amended by T.D. ATF–293, 55 FR 3722, Feb. 5, 1990; T.D. ATF–400, 63 FR 45003, Aug. 24, 1998]

§ 555.203 Types of magazines.

For purposes of this part, there are five types of magazines. These types, together with the classes of explosive materials, as defined in § 555.202, which will be stored in them, are as follows:

(a) **Type 1 magazines.** Permanent magazines for the storage of high explosives, subject to the limitations prescribed by §§ 555.206 and 555.213. Other classes of explosive materials may also be stored in type 1 magazines.

(b) **Type 2 magazines.** Mobile and portable indoor and outdoor magazines for the storage of high explosives, subject to the limitations prescribed by §§§ 555.206, 555.208(b), and 555.213. Other classes of explosive materials may also be stored in type 2 magazines.

(c) **Type 3 magazines.** Portable outdoor magazines for the temporary storage of high explosives while attended (for example, a "day-box"), subject to the limitations prescribed by §§ 555.206 and 555.213. Other classes of explosives materials may also be stored in type 3 magazines.

(d) **Type 4 magazines.** Magazines for the storage of low explosives, subject to the limitations prescribed by §§§ 555.206(b), 555.210(b), and 555.213. Blasting agents may be stored in type 4 magazines, subject to the limitations prescribed by §§§ 555.206(c), 555.211(b), and 555.213. Detonators that will not mass detonate may also be stored in type 4 magazines, subject to the limitations prescribed by §§§ 555.206(a), 555.210(b), and 555.213.

(e) **Type 5 magazines.** Magazines for the storage of blasting agents, subject to the limitations prescribed by §§§ 555.206(c), 555.211(b), and 555.213.

§ 555.204 Inspection of magazines.

Any person storing explosive materials shall inspect his magazines at least every seven days. This inspection need not be an inventory, but must be sufficient to determine whether there has been unauthorized entry or attempted entry into the magazines, or unauthorized removal of the contents of the magazines.

§ 555.205 Movement of explosive materials.

All explosive materials must be kept in locked magazines meeting the standards in this subpart unless they are:

(a) In the process of manufacture;

(b) Being physically handled in the operating process of a licensee or user;

(c) Being used; or

(d) Being transported to a place of storage or use by a licensee or permittee or by a person who has lawfully acquired explosive materials under § 555.106.

§ 555.206 Location of magazines.

(a) Outdoor magazines in which high explosives are stored must be located no closer to inhabited buildings, passenger railways, public highways, or other magazines in which high explosives are stored, than the minimum distances specified in the table of distances for storage of explosive materials in § 555.218.

(b) Outdoor magazines in which low explosives are stored must be located no closer to inhabited buildings, passenger railways, public highways, or other magazines in which explosive materials are stored, than the minimum distances specified in the table of distances for storage of low explosives in § 555.219, except that the table of distances in § 555.224 shall apply to the storage of display fireworks. The distances shown in § 555.219 may not be reduced by the presence of barricades.

(c) **(1)** Outdoor magazines in which blasting agents in quantities of more than 50 pounds are stored must be located no closer to inhabited buildings, passenger railways, or public highways than the minimum distances specified in the table of distances for storage of explosive materials in § 555.218.

(2) Ammonium nitrate and magazines in which blasting agents are stored must be located no closer to magazines in which high explosives or other blasting agents are stored than the minimum distances specified in the table of distances for the separation of ammonium nitrate and blasting agents in § 555.220. However, the minimum distances for magazines in which explosives and blasting agents are stored from inhabited buildings, etc., may not be less than the distances specified in the table of distances for storage of explosives materials in § 555.218.

[T.D. ATF–87, 46 FR 40384, Aug. 7, 1981, as amended by T.D. ATF–293, 55 FR 3722, Feb. 5, 1990; T.D. ATF–400, 63 FR 45003, Aug. 24, 1998]

§ 555.207 Construction of type 1 magazines.

A type 1 magazine is a permanent structure: a building, an igloo or "Army-type structure", a tunnel, or a dugout. It is to be bullet-resistant, fire-resistant, weather-resistant, theft-resistant, and ventilated.

(a) **Buildings.** All building type magazines are to be constructed of masonry, wood, metal, or a combination of these materials, and have no openings except for entrances and ventilation. The ground around building magazines must slope away for drainage or other adequate drainage provided.

(1) **Masonry wall construction.** Masonry wall construction is to consist of brick, concrete, tile, cement block, or cinder block and be not less than 6" in thickness. Hollow masonry units used in construction must have all hollow spaces filled with well-tamped, coarse, dry sand or weak concrete (at least a mixture of one part cement and eight parts of sand with enough water to dampen the mixture while tamping in place). Interior walls are to be constructed of, or covered with, a nonsparking material.

(2) **Fabricated metal wall construction.** Metal wall construction is to consist of sectional sheets of steel or aluminum not less than number 14-gauge, securely fastened to a metal framework. Metal wall construction is either lined inside with brick, solid cement blocks, hardwood not less than four inches thick, or will have at least a six inch sand fill between interior and exterior walls. Interior walls are to be constructed of, or covered with, a nonsparking material.

(3) **Wood frame wall construction.** The exterior of outer wood walls is to be covered with iron or aluminum not less than number 26-gauge. An inner wall of, or covered with nonsparking material will be constructed so as to provide a space of not less than six inches between the outer and inner walls. The space is to be filled with coarse, dry sand or weak concrete.

(4) **Floors.** Floors are to be constructed of, or covered with, a nonsparking material and shall be strong enough to bear the weight of the maximum quantity to be stored. Use of pallets covered with a nonsparking material is considered equivalent to a floor constructed of or covered with a nonsparking material.

(5) **Foundations.** Foundations are to be constructed of brick, concrete, cement block, stone, or wood posts. If piers or posts are used, in lieu of a continuous foundation, the space under the buildings is to be enclosed with metal.

(6) **Roof.** Except for buildings with fabricated metal roofs, the outer roof is to be covered with no less than number 26-guage iron or aluminum, fastened to at least ⅞" sheathing.

(7) **Bullet-resistant ceilings or roofs.** Where it is possible for a bullet to be fired directly through the roof and into the magazine at such an angle that the bullet would strike the explosives within, the magazine is to be protected by one of the following methods:

(i) A sand tray lined with a layer of building paper, plastic, or other nonporous material, and filled with not less than four inches of coarse, dry sand, and located at the tops of inner walls covering the entire ceiling area, except that portion necessary for ventilation.

(ii) A fabricated metal roof constructed of $\frac{3}{16}$" plate steel lined with four inches of hardwood. (For each additional $\frac{1}{16}$" of plate steel, the hardwood lining may be decreased one inch.)

(8) Doors. All doors are to be constructed of not less than $\frac{1}{4}$" plate steel and lined with at least two inches of hardwood. Hinges and hasps are to be attached to the doors by welding, riveting or bolting (nuts on inside of door). They are to be installed in such a manner that the hinges and hasps cannot be removed when the doors are closed and locked.

(9) Locks. Each door is to be equipped with (i) two mortise locks; (ii) two padlock fastened in separate hasps and staples; (iii) a combination of a mortise lock and a padlock; (iv) a mortise lock that requires two keys to open; or (v) a three-point lock. Padlocks must have at least five tumblers and a casehardened shackle of at least $\frac{3}{8}$" diameter. Padlocks must be protected with not less than $\frac{1}{4}$" steel hoods constructed so as to prevent sawing or lever action on the locks, hasps, and staples. These requirements do not apply to magazine doors that are adequately secured on the inside by means of a bolt, lock, or bar that cannot be actuated from the outside.

(10) Ventilation. Ventilation is to be provided to prevent dampness and heating of stored explosive materials. Ventilation openings must be screened to prevent the entrance of sparks. Ventilation openings in side walls and foundations must be offset or shielded for bullet-resistant purposes. Magazines having foundation and roof ventilators with the air circulating between the side walls and the floors and between the side walls and the ceiling must have a wooden lattice lining or equivalent to prevent the packages of explosive materials from being stacked against the side walls and blocking the air circulation.

(11) Exposed metal. No sparking material is to be exposed to contact with the stored explosive materials. All ferrous metal nails in the floor and side walls, which might be exposed to contact with explosive materials, must be blind nailed, countersunk, or covered with a nonsparking lattice work or other nonsparking material.

(b) Igloos, "Army-type structures", tunnels, and dugouts. Igloo, "Army-type structure", tunnel, and dugout magazines are to be constructed of reinforced concrete, masonry, metal, or a combination of these materials. They must have an earthmound covering of not less than 24" on the top, sides and rear unless the magazine meets the requirements of paragraph (a)(7) of this section. Interior walls and floors must be constructed of, or covered with, a nonsparking material. Magazines of this type are also to be constructed in conformity with the requirements of paragraph (a)(4) and paragraphs (a)(8) through (11) of this section.

§ 555.208 Construction of type 2 magazines.

A type 2 magazine is a box, trailer, semitrailer, or other mobile facility.

(a) Outdoor magazines—

(1) General. Outdoor magazines are to be bullet-resistant, fire-resistant, weather-resistant, theft-resistant, and ventilated. They are to be supported to prevent direct contact with the ground and, if less than one cubic yard in size, must be securely fastened to a fixed object. The ground around outdoor magazines must slope away for drainage or other adequate drainage provided. When unattended, vehicular magazines must have wheels removed or otherwise effectively immobilized by kingpin locking devices or other methods approved by the Director.

(2) Exterior construction. The exterior and doors are to be constructed of not less than $\frac{1}{4}$" steel and lined with at least two inches of hardwood. Magazines with top openings will have lids with water-resistant seals or which overlap the sides by at least one inch when in a closed position.

(3) Hinges and hasps. Hinges and hasps are to be attached to doors by welding, riveting, or bolting (nuts on inside of door). Hinges and hasps must be installed so that they cannot be removed when the doors are closed and locked.

(4) Locks. Each door is to be equipped with (i) two mortise locks; (ii) two padlocks fastened in separate hasps and staples; (iii) a combination of a mortise lock and a padlock; (iv) a mortise lock that requires two keys to open; or (v) a three-point lock. Padlocks must have at least five tumblers and a case-hardened shackle of at least $\frac{3}{8}$" diameter. Padlocks must be protected with not less than $\frac{1}{4}$" steel hoods constructed so as to prevent sawing or lever action on the locks, hasps, and staples. These requirements do not apply to magazine doors that are adequately secured on the inside by means of a bolt, lock, or bar that cannot be actuated from the outside.

(b) Indoor magazines—

(1) General. Indoor magazines are to be fire-resistant and theft-resistant. They need not be bullet-resistant and weather-resistant if the buildings in which they are stored provide protection from the weather and from bullet penetration. No indoor magazine is to be located in a residence or dwelling. The indoor storage of high explosives must not exceed a quantity of 50 pounds. More than one indoor magazine may be located in the same building if the total quantity of explosive materials stored does not exceed 50 pounds. Detonators must be stored in a separate magazine (except as provided in § 555.213) and the total quantity of detonators must not exceed 5,000.

(2) Exterior construction. Indoor magazines are to be constructed of wood or metal according to one of the following specifications:

(i) Wood indoor magazines are to have sides, bottoms and doors constructed of at least two inches of hardwood and are to be well braced at the corners. They are to be covered with sheet metal of not less than number 26-gauge (.0179 inches).

Nails exposed to the interior of magazines must be countersunk.

(ii) Metal indoor magazines are to have sides, bottoms and doors constructed of not less than number 12-gauge (.1046 inches) metal and be lined inside with a nonsparking material. Edges of metal covers must overlap sides at least one inch.

(3) **Hinges and hasps.** Hinges and hasps are to be attached to doors by welding, riveting, or bolting (nuts on inside of door). Hinges and hasps must be installed so that they cannot be removed when the doors are closed and locked.

(4) **Locks.** Each door is to be equipped with (i) two mortise locks; (ii) two padlocks fastened in separate hasps and staples; (iii) a combination of a mortise lock and a padlock; (iv) a mortise lock that requires two keys to open; or (v) a three-point lock. Padlocks must have at least five tumblers and a case-hardened shackle of at least ⅜" diameter. Padlocks must be protected with not less than ¼" steel hoods constructed so as to prevent sawing or lever action on the locks, hasps, and staples. Indoor magazines located in secure rooms that are locked as provided in this subparagraph may have each door locked with one steel padlock (which need not be protected by a steel hood) having at least five tumblers and a case-hardened shackle of at least ⅜" diameter, if the door hinges and lock hasp are securely fastened to the magazine.

These requirements do not apply to magazine doors that are adequately secured on the inside by means of a bolt, lock, or bar that cannot be actuated from the outside.

(c) **Detonator boxes.** Magazines for detonators in quantities of 100 or less are to have sides, bottoms and doors constructed of not less than number 12-gauge (.1046 inches) metal and lined with a nonsparking material. Hinges and hasps must be attached so they cannot be removed from the outside. One steel padlock (which need not be protected by a steel hood) having at least five tumblers and a case-hardened shackle of at least ⅜" diameter is sufficient for locking purposes.

§555.209 Construction of type 3 magazines.

A type 3 magazine is a "day-box" or other portable magazine. It must be fire-resistant, weather-resistant, and theft-resistant. A type 3 magazine is to be constructed of not less than number 12-gauge (.1046 inches) steel, lined with at least either ½" plywood or ½" Masonite-type hardboard. Doors must overlap sides by at least one inch. Hinges and hasps are to be attached by welding, riveting or bolting (nuts on inside). One steel padlock (which need not be protected by a steel hood) having at least five tumblers and a case-hardened shackle of at least ⅜" diameter is sufficient for locking purposes. Explosive materials are not to be left unattended in type 3 magazines and must be removed to type 1 or 2 magazines for unattended storage.

§555.210 Construction of type 4 magazines.

A type 4 magazine is a building, igloo or "Army-type structure", tunnel, dugout, box, trailer, or a semitrailer or other mobile magazine.

(a) **Outdoor magazines**—

(1) **General.** Outdoor magazines are to be fire-resistant, weather-resistant, and theft-resistant. The ground around outdoor magazines must slope away for drainage or other adequate drainage be provided. When unattended, vehicular magazines must have wheels removed or otherwise be effectively immobilized by kingpin locking devices or other methods approved by the Director.

(2) **Construction.** Outdoor magazines are to be constructed of masonry, metal-covered wood, fabricated metal, or a combination of these materials. Foundations are to be constructed of brick, concrete, cement block, stone, or metal or wood posts. If piers or posts are used, in lieu of a continuous foundation, the space under the building is to be enclosed with fire-resistant material. The walls and floors are to be constructed of, or covered with, a nonsparking material or lattice work. The doors must be metal or solid wood covered with metal.

(3) **Hinges and hasps.** Hinges and hasps are to be attached to doors by welding, riveting, or bolting (nuts on inside of door). Hinges and hasps must be installed so that they cannot be removed when the doors are closed and locked.

(4) **Locks.** Each door is to be equipped with (i) two mortise locks; (ii) two padlocks fastened in separate hasps and staples; (iii) a combination of a mortise lock and a padlock; (iv) a mortise lock that requires two keys to open; or (v) a three-point lock. Padlocks must have at least five tumblers and case-hardened shackle of at least ⅜" diameter. Padlocks must be protected with not less than ¼" steel hoods constructed so as to prevent sawing or lever action on the locks, hasps, and staples. These requirements do not apply to magazine doors that are adequately secured on the inside by means of a bolt, lock, or bar that cannot be actuated from the outside.

(b) **Indoor magazine**—

(1) **General.** Indoor magazines are to be fire-resistant and theft-resistant. They need not be weather-resistant if the buildings in which they are stored provide protection from the weather. No indoor magazine is to be located in a residence or dwelling. The indoor storage of low explosives must not exceed a quantity of 50 pounds. More than one indoor magazine may be located in the same building if the total quantity of explosive materials stored does not exceed 50 pounds. Detonators that will not mass detonate must be stored in a separate magazine and the total number of electric detonators must not exceed 5,000.

(2) **Construction.** Indoor magazines are to be constructed of masonry, metal-covered wood, fabricated metal, or a combination of these materials. The walls and floors are to be constructed of, or covered with, a nonsparking material. The doors must be metal or solid wood covered with metal.

(3) Hinges and hasps. Hinges and hasps are to be attached to doors by welding, riveting, or bolting (nuts on inside of door). Hinges and hasps must be installed so that they cannot be removed when the doors are closed and locked.

(4) Locks. Each door is to be equipped with (i) two mortise locks; (ii) two padlocks fastened in separate hasps and staples; (iii) a combination of a mortise lock and padlock; (iv) a mortise lock that requires two keys to open; or (v) a three-point lock. Padlocks must have at least five tumblers and a case-hardened shackle of at least ⅜" diameter. Padlocks must be protected with not less than ¼" steel hoods constructed so as to prevent sawing or lever action on the locks, hasps, and staples. Indoor magazines located in secure rooms that are locked as provided in this subparagraph may have each door locked with one steel padlock (which need not be protected by a steel hood) having at least five tumblers and a case-hardened shackle of at least ⅜" diameter, if the door hinges and lock hasp are securely fastened to the magazine. These requirements do not apply to magazine doors that are adequately secured on the inside by means of a bolt, lock, or bar that cannot be actuated from the outside.

§ 555.211 Construction of type 5 magazines.

A type 5 magazine is a building, igloo or "Army-type structure", tunnel, dugout, bin, box, trailer, or a semitrailer or other mobile facility.

(a) Outdoor magazines—

(1) General. Outdoor magazines are to be weather-resistant and theft-resistant. The ground around magazines must slope away for drainage or other adequate drainage be provided. When unattended, vehicular magazines must have wheels removed or otherwise be effectively immobilized by kingpin locking devices or other methods approved by the Director.

(2) Construction. The doors are to be constructed of solid wood or metal.

(3) Hinges and hasps. Hinges and hasps are to be attached to doors by welding, riveting, or bolting (nuts on inside of door). Hinges and hasps must be installed so that they cannot be removed when the doors are closed and locked.

(4) Locks. Each door is to be equipped with (i) two mortise locks; (ii) two padlocks fastened in separate hasps and staples; (iii) a combination of a mortise lock and a padlock; (iv) a mortise lock that requires two keys to open; or (v) a three-point lock. Padlocks must have at least five tumblers and a case-hardened shackle of at least ⅜" diameter. Padlocks must be protected with not less than ¼" steel hoods constructed so as to prevent sawing or lever action on the locks, hasps, and staples. Trailers, semitrailers, and similar vehicular magazines may, for each door, be locked with one steel padlock (which need not be protected by a steel hood) having at least five tumblers and a case-hardened shackle of at least ⅜" diameter, if the door hinges and lock hasp are securely fastened to the magazine and to the door frame.

These requirements do not apply to magazine doors that are adequately secured on the inside by means of a bolt, lock, or bar that cannot be actuated from the outside.

(5) Placards. The placards required by Department of Transportation regulations at 49 CFR part 172, subpart F, for the transportation of blasting agents shall be displayed on all magazines.

(b) Indoor magazines—

(1) General. Indoor magazines are to be theft-resistant. They need not be weather-resistant if the buildings in which they are stored provide protection from the weather. No indoor magazine is to be located in a residence or dwelling. Indoor magazines containing quantities of blasting agents in excess of 50 pounds are subject to the requirements of § 555.206 of this subpart.

(2) Construction. The doors are to be constructed of wood or metal.

(3) Hinges and hasps. Hinges and hasps are to be attached to doors by welding, riveting, or bolting (nuts on inside). Hinges and hasps must be installed so that they cannot be removed when the doors are closed and locked.

(4) Locks. Each door is to be equipped with (i) two mortise locks; (ii) two padlocks fastened in separate hasps and staples; (iii) a combination of a mortise lock and a padlock; (iv) a mortise lock that requires two keys to open; or (v) a three-point lock. Padlocks must have at least five tumblers and a case-hardened shackle of at least ⅜" diameter. Padlocks must be protected with not less than ¼" steel hoods constructed so as to prevent sawing or lever action on the locks, hasps, and staples. Indoor magazines located in secure rooms that are locked as provided in this subparagraph may have each door locked with one steel padlock (which need not be protected by a steel hood) having at least five tumblers and a case-hardened shackle of at least ⅜" diameter, if the door hinges and lock hasps are securely fastened to the magazine and to the door frame. These requirements do not apply to magazine doors that are adequately secured on the inside by means of a bolt, lock, or bar that cannot be actuated from the outside.

[T.D. ATF–87, 46 FR 40384, Aug. 7, 1981, as amended by T.D. ATF–298, 55 FR 21863, May 30, 1990]

§ 555.212 Smoking and open flames.

Smoking, matches, open flames, and spark producing devices are not permitted:

(a) In any magazine;

(b) Within 50 feet of any outdoor magazine; or

(c) Within any room containing an indoor magazine.

§ 555.213 Quantity and storage restrictions.

(a) Explosive materials in excess of 300,000 pounds or detonators in excess of 20 million are not to be stored in one magazine unless approved by the Director.

(b) Detonators are not to be stored in the same magazine with other explosive materials, except under the following circumstances:

    (1) In a type 4 magazine, detonators that will not mass detonate may be stored with electric squibs, safety fuse, shock tube, igniters, and igniter cord.

    (2) In a type 1 or type 2 magazine, detonators may be stored with delay devices and any of the items listed in paragraph (b)(1) of this section.

[T.D. ATF–487, 68 FR 3748, Jan. 24, 2003, as amended by ATF 15F, 75 FR 3163, Jan. 20, 2010]

§ 555.214 Storage within types 1, 2, 3, and 4 magazines.

(a) Explosive materials within a magazine are not to be placed directly against interior walls and must be stored so as not to interfere with ventilation. To prevent contact of stored explosive materials with walls, a nonsparking lattice work or other nonsparking material may be used.

(b) Containers of explosive materials are to be stored so that marks are visible. Stocks of explosive materials are to be stored so they can be easily counted and checked upon inspection.

(c) Except with respect to fiberboard or other nonmetal containers, containers of explosive materials are not to be unpacked or repacked inside a magazine or within 50 feet of a magazine, and must not be unpacked or repacked close to other explosive materials. Containers of explosive materials must be closed while being stored.

(d) Tools used for opening or closing containers of explosive materials are to be of nonsparking materials, except that metal slitters may be used for opening fiberboard containers. A wood wedge and a fiber, rubber, or wooden mallet are to be used for opening or closing wood containers of explosive materials. Metal tools other than nonsparking transfer conveyors are not to be stored in any magazine containing high explosives.

§ 555.215 Housekeeping.

Magazines are to be kept clean, dry, and free of grit, paper, empty packages and containers, and rubbish. Floors are to be regularly swept. Brooms and other utensils used in the cleaning and maintenance of magazines must have no spark-producing metal parts, and may be kept in magazines. Floors stained by leakage from explosive materials are to be cleaned according to instructions of the explosives manufacturer. When any explosive material has deteriorated it is to be destroyed in accordance with the advice or instructions of the manufacturer. The area surrounding magazines is to be kept clear of rubbish, brush, dry grass, or trees (except live trees more than 10 feet tall), for not less than 25 feet in all directions. Volatile materials are to be kept a distance of not less than 50 feet from outdoor magazines. Living foliage which is used to stabilize the earthen covering of a magazine need not be removed.

§ 555.216 Repair of magazines.

Before repairing the interior of magazines, all explosive materials are to be removed and the interior cleaned. Before repairing the exterior of magazines, all explosive materials must be removed if there exists any possibility that repairs may produce sparks or flame. Explosive materials removed from magazines under repair must be (a) placed in other magazines appropriate for the storage of those explosive materials under this subpart, or (b) placed a safe distance from the magazines under repair where they are to be properly guarded and protected until the repairs have been completed.

§ 555.217 Lighting.

(a) Battery-activated safety lights or battery-activated safety lanterns may be used in explosives storage magazines.

(b) Electric lighting used in any explosives storage magazine must meet the standards prescribed by the "National Electrical Code," (National Fire Protection Association, NFPA 70–81), for the conditions present in the magazine at any time. All electrical switches are to be located outside of the magazine and also meet the standards prescribed by the National Electrical Code.

(c) Copies of invoices, work orders or similar documents which indicate the lighting complies with the National Electrical Code must be available for inspection by ATF officers.

§555.218 Table of distances for storage of explosive materials.

Quantity of Explosives		Distances in feet							
Pounds over	Pounds not over	Inhabited buildings		Public highways with traffic volume 3000 or fewer vehicles/day		Passenger railways-public highways with traffic volume more than 3,000 vehicles/day		Separation of magazines	
		Barricaded	Unbarricaded	Barricaded	Unbarricaded	Barricaded	Unbarricaded	Barricaded	Unbarricaded
0	5	70	140	30	60	51	102	6	12
5	10	90	180	35	70	64	128	8	16
10	20	110	220	45	90	81	162	10	20
20	30	125	250	50	100	93	186	11	22
30	40	140	280	55	110	103	206	12	24
40	50	150	300	60	120	110	220	14	28
50	75	170	340	70	140	127	254	15	30
75	100	190	380	75	150	139	278	16	32
100	125	200	400	80	160	150	300	18	36
125	150	215	430	85	170	159	318	19	38
150	200	235	470	95	190	175	350	21	42
200	250	255	510	105	210	189	378	23	46
250	300	270	540	110	220	201	402	24	48
300	400	295	590	120	240	221	442	27	54
400	500	320	640	130	260	238	476	29	58
500	600	340	680	135	270	253	506	31	62
600	700	355	710	145	290	266	532	32	64
700	800	375	750	150	300	278	556	33	66
800	900	390	780	155	310	289	578	35	70
900	1,000	400	800	160	320	300	600	36	72
1,000	1,200	425	850	165	330	318	636	39	78
1,200	1,400	450	900	170	340	336	672	41	82
1,400	1,600	470	940	175	350	351	702	43	86
1,600	1,800	490	980	180	360	366	732	44	88
1,800	2,000	505	1,010	185	370	378	756	45	90
2,000	2,500	545	1,090	190	380	408	816	49	98
2,500	3,000	580	1,160	195	390	432	864	52	104
3,000	4,000	635	1,270	210	420	474	948	58	116
4,000	5,000	685	1,370	225	450	513	1,026	61	122
5,000	6,000	730	1,460	235	470	546	1,092	65	130
6,000	7,000	770	1,540	245	490	573	1,146	68	136
7,000	8,000	800	1,600	250	500	600	1,200	72	144
8,000	9,000	835	1,670	255	510	624	1,248	75	150
9,000	10,000	865	1,730	260	520	645	1,290	78	156
10,000	12,000	875	1,750	270	540	687	1,374	82	164
12,000	14,000	885	1,770	275	550	723	1,446	87	174
14,000	16,000	900	1,800	280	560	756	1,512	90	180
16,000	18,000	940	1,880	285	570	786	1,572	94	188
18,000	20,000	975	1,950	290	580	813	1,626	98	196
20,000	25,000	1,055	2,000	315	630	876	1,752	105	210
25,000	30,000	1,130	2,000	340	680	933	1,866	112	224
30,000	35,000	1,205	2,000	360	720	981	1,962	119	238
35,000	40,000	1,275	2,000	380	760	1,026	2,000	124	248
40,000	45,000	1,340	2,000	400	800	1,068	2,000	129	258
45,000	50,000	1,400	2,000	420	840	1,104	2,000	135	270
50,000	55,000	1,460	2,000	440	880	1,140	2,000	140	280
55,000	60,000	1,515	2,000	455	910	1,173	2,000	145	290
60,000	65,000	1,565	2,000	470	940	1,206	2,000	150	300
65,000	70,000	1,610	2,000	485	970	1,236	2,000	155	310
70,000	75,000	1,655	2,000	500	1,000	1,263	2,000	160	320
75,000	80,000	1,695	2,000	510	1,020	1,293	2,000	165	330
80,000	85,000	1,730	2,000	520	1,040	1,317	2,000	170	340
85,000	90,000	1,760	2,000	530	1,060	1,344	2,000	175	350
90,000	95,000	1,790	2,000	540	1,080	1,368	2,000	180	360
95,000	100,000	1,815	2,000	545	1,090	1,392	2,000	185	370
100,000	110,000	1,835	2,000	550	1,100	1,437	2,000	195	390
110,000	120,000	1,855	2,000	555	1,110	1,479	2,000	205	410
120,000	130,000	1,875	2,000	560	1,120	1,521	2,000	215	430
130,000	140,000	1,890	2,000	565	1,130	1,557	2,000	225	450
140,000	150,000	1,900	2,000	570	1,140	1,593	2,000	235	470
150,000	160,000	1,935	2,000	580	1,160	1,629	2,000	245	490
160,000	170,000	1,965	2,000	590	1,180	1,662	2,000	255	510
170,000	180,000	1,990	2,000	600	1,200	1,695	2,000	265	530
180,000	190,000	2,010	2,010	605	1,210	1,725	2,000	275	550
190,000	200,000	2,030	2,030	610	1,220	1,755	2,000	285	570
200,000	210,000	2,055	2,055	620	1,240	1,782	2,000	295	590
210,000	230,000	2,100	2,100	635	1,270	1,836	2,000	315	630
230,000	250,000	2,155	2,155	650	1,300	1,890	2,000	335	670
250,000	275,000	2,215	2,215	670	1,340	1,950	2,000	360	720
275,000	300,000	2,275	2,275	690	1,380	2,000	2,000	385	770

Table: American Table of Distances for Storage of Explosives (December 1910), as Revised and Approved by the Institute of Makers of Explosives—July, 1991.

Notes to the Table of Distances for Storage of Explosives

(1) Terms found in the table of distances for storage of explosive materials are defined in § 555.11.

(2) When two or more storage magazines are located on the same property, each magazine must comply with the minimum distances specified from inhabited buildings, railways, and highways, and, in addition, they should be separated from each other by not less than the distances shown for "Separation of Magazines," except that the quantity of explosives contained in cap magazines shall govern in regard to the spacing of said cap magazines from magazines containing other explosives. If any two or more magazines are separated from each other by less than the specified "Separation of Magazines" distances, then such two or more magazines, as a group, must be considered as one magazine, and the total quantity of explosives stored in such group must be treated as if stored in a single magazine located on the site of any magazine of the group, and must comply with the minimum of distances specified from other magazines, inhabited buildings, railways, and highways.

(3) All types of blasting caps in strengths through No. 8 cap should be rated at 1 ½ lbs. (1.5 lbs.) of explosives per 1,000 caps. For strengths higher than No. 8 cap, consult the manufacturer.

(4) For quantity and distance purposes, detonating cord of 50 or 60 grains per foot should be calculated as equivalent to 9 lbs. of high explosives per 1,000 feet. Heavier or lighter core loads should be rated proportionately.

[T.D. ATF–87, 46 FR 40384, Aug. 7, 1981, as amended by T.D. ATF–400, 63 FR 45003, Aug. 24, 1998; T.D. ATF–446, 66 FR 16602, Mar. 27, 2001; T.D. ATF–446a, 66 FR 19089, Apr. 13, 2001]

§ 555.219 Table of distances for storage of low explosives.

Pounds		From Inhabited building distance (feet)	From public railroad and highway distance (feet)	From above ground magazine (feet)
Over	Not over			
0	1,000	75	75	50
1,000	5,000	115	115	75
5,000	10,000	150	150	100
10,000	20,000	190	190	125
20,000	30,000	215	215	145
30,000	40,000	235	235	155
40,000	50,000	250	250	165
50,000	60,000	260	260	175
60,000	70,000	270	270	185
70,000	80,000	280	280	190
80,000	90,000	295	295	195
90,000	100,000	300	300	200
100,000	200,000	375	375	250
200,000	300,000	450	450	300

Table: Department of Defense Ammunition and Explosives Standards, Table 5–4.1 Extract; 4145.27 M, March 1969

§ 555.220 Table of separation distances of ammonium nitrate and blasting agents from explosives or blasting agents.

Donor weight (pounds)		Minimum separation distance of acceptor from donor when barricaded (feet)		Minimum thickness of artificial barricades (inches)
Over	Not over	Ammonium nitrate	Blasting agent	
0	100	3	11	12
100	300	4	14	12
300	600	5	18	12
600	1,000	6	22	12
1,000	1,600	7	25	12
1,600	2,000	8	29	12
2,000	3,000	9	32	15
3,000	4,000	10	36	15
4,000	6,000	11	40	15
6,000	8,000	12	43	20
8,000	10,000	13	47	20
10,000	12,000	14	50	20
12,000	16,000	15	54	25
16,000	20,000	16	58	25
20,000	25,000	18	65	25
25,000	30,000	19	68	30
30,000	35,000	20	72	30
35,000	40,000	21	76	30
40,000	45,000	22	79	35
45,000	50,000	23	83	35
50,000	55,000	24	86	35
55,000	60,000	25	90	35
60,000	70,000	26	94	40
70,000	80,000	28	101	40
80,000	90,000	30	108	40
90,000	100,000	32	115	40
100,000	120,000	34	122	50
120,000	140,000	37	133	50
140,000	160,000	40	144	50
160,000	180,000	44	158	50
180,000	200,000	48	173	50
200,000	220,000	52	187	60
220,000	250,000	56	202	60
250,000	275,000	60	216	60
275,000	300,000	64	230	60

Table: National Fire Protection Association (NFPA) Official Standard No. 492, 1968

Notes of Table of Separation Distances of Ammonium Nitrate and Blasting Agents From Explosives or Blasting Agents

(1) This table specifies separation distances to prevent explosion of ammonium nitrate and ammonium nitrate-based blasting agents by propagation from nearby stores of high explosives or blasting agents referred to in the table as the "donor." Ammonium nitrate, by itself, is not considered to be a donor when applying this table. Ammonium nitrate, ammonium nitrate-fuel oil or combinations thereof are acceptors. If stores of ammonium nitrate are located within the sympathetic detonation distance of explosives or blasting agents, one-half the mass of the ammonium nitrate is to be included in the mass of the donor.

(2) When the ammonium nitrate and/or blasting agent is not barricaded, the distances shown in the table must be multiplied by six. These distances allow for the possibility of high velocity metal fragments from mixers, hoppers, truck bodies, sheet metal structures, metal containers, and the like which may enclose the "donor." Where explosives storage is in bullet-resistant magazines or where the storage is protected by a bullet-resistant wall, distances and barricade thicknesses in excess of those prescribed in the table in § 555.218 are not required.

(3) These distances apply to ammonium nitrate that passes the insensitivity test prescribed in the definition of ammonium nitrate fertilizer issued by the Fertilizer Institute. [1] Ammonium nitrate failing to pass the test must be stored at separation distances in accordance with the table in § 555.218.

[1] Definition and Test Procedures for Ammonium Nitrate Fertilizer, Fertilizer Institute 425 Third Street, SW, Suite 950, Washington, D.C. 20024.

(4) These distances apply to blasting agents which pass the insensitivity test prescribed in regulations of the U.S. Department of Transportation (49 CFR part 173).

(5) Earth or sand dikes, or enclosures filled with the prescribed minimum thickness of earth or sand are acceptable artificial barricades. Natural barricades, such as hills or timber of

sufficient density that the surrounding exposures which require protection cannot be seen from the "donor" when the trees are bare of leaves, are also acceptable.

(6) For determining the distances to be maintained from inhabited buildings, passenger railways, and public highways, use the table in § 555.218.

§ 555.221 Requirements for display fireworks, pyrotechnic compositions, and explosive materials used in assembling fireworks or articles pyrotechnic.

(a) Display fireworks, pyrotechnic compositions, and explosive materials used to assemble fireworks and articles pyrotechnic shall be stored at all times as required by this Subpart unless they are in the process of manufacture, assembly, packaging, or are being transported.

(b) No more than 500 pounds (227 kg) of pyrotechnic compositions or explosive materials are permitted at one time in any fireworks mixing building, any building or area in which the pyrotechnic compositions or explosive materials are pressed or otherwise prepared for finishing or assembly, or any finishing or assembly building. All pyrotechnic compositions or explosive materials not in immediate use will be stored in covered, non-ferrous containers.

(c) The maximum quantity of flash powder permitted in any fireworks process building is 10 pounds (4.5 kg).

(d) All dry explosive powders and mixtures, partially assembled display fireworks, and finished display fireworks shall be removed from fireworks process buildings at the conclusion of a day's operations and placed in approved magazines.

[T.D. ATF–293, 55 FR 3722, Feb. 5, 1990, as amended by T.D. ATF–400, 63 FR 45004, Aug. 24, 1998]

§ 555.222 Table of distances between fireworks process buildings and between fireworks process and fireworks nonprocess buildings.

Net weight of fireworks[1] (pounds)	Display fireworks[2] (feet)	Consumer fireworks[3] (feet)
0-100	57	37
101-200	69	37
201-300	77	37
301-400	85	37
401-500	91	37
Above 500	Not permitted[4] [5]	Not permitted[4] [5]

[1] Net weight is the weight of all pyrotechnic compositions, and explosive materials and fuse only.

[2] The distances in this column apply only with natural or artificial barricades. If such barricades are not used, the distances must be doubled.

[3] While consumer fireworks or articles pyrotechnic in a finished state are not subject to regulation, explosive materials used to manufacture or assemble such fireworks or articles are subject to regulation. Thus, fireworks process buildings where consumer fireworks or articles pyrotechnic are being processed shall meet these requirements.

[4] A maximum of 500 pounds of in-process pyrotechnic compositions, either loose or in partially-assembled fireworks, is permitted in any fireworks process building. Finished display fireworks may not be stored in a fireworks process building.

[5] A maximum of 10 pounds of flash powder, either in loose form or in assembled units, is permitted in any fireworks process building. Quantities in excess of 10 pounds must be kept in an approved magazine.

[T.D. ATF–293, 55 FR 3723, Feb. 5, 1990, as amended by T.D. ATF–400, 63 FR 45004, Aug. 24, 1998]

§ 555.223 Table of distances between fireworks process buildings and other specified areas.

Net weight of fireworks[1] (pounds)	Display fireworks[1] (feet)	Consumer fireworks[2] (feet)
0-100	200	25
101-200	200	50
201-300	200	50
301-400	200	50
401-500	200	50
Above 500	Not permitted	Not permitted

Distance from Passenger Railways, Public Highways, Fireworks Plant Buildings used to Store Consumer Fireworks and Articles Pyrotechnic, Magazines and Fireworks Shipping Buildings, and Inhabited Buildings.[3,4,5]

[1] Net weight is the weight of all pyrotechnic compositions, and explosive materials and fuse only.

[2] While consumer fireworks or articles pyrotechnic in a finished state are not subject to regulation, explosive materials used to manufacture or assemble such fireworks or articles are subject to regulation. Thus, fireworks process buildings where consumer fireworks or articles pyrotechnic are being processed shall meet these requirements.

[3] This table does not apply to the separation distances between fireworks process buildings (see § 555.222) and between magazines (see §§ 555.218 and 555.224).

[4] The distances in this table apply with or without artificial or natural barricades or screen barricades. However, the use of barricades is highly recommended.

[5] No work of any kind, except to place or move items other than explosive materials from storage, shall be conducted in any building designated as a warehouse. A fireworks plant warehouse is not subject to § 555.222 or this section, tables of distances.

[T.D ATF–293, 55 FR 3723, Feb. 5, 1990, as amended by T.D. ATF–400, 63 FR 45004, Aug. 24, 1998]

§ 555.224 Table of distances for the storage of display fireworks (except bulk salutes).

Net weight of fireworks[1] (pounds)	Distance between magazine and inhabited building, passenger railway, or public highway[3] [4] (feet)	Distance between magazines[2] [3] (feet)
0-1000	150	100
1,001-5,000	230	150
5,001-10,000	300	200
Above 10,000	Use Table § 555.218	

[1] Net weight is the weight of all pyrotechnic compositions, and explosive materials and fuse only.

[2] For the purposes of applying this table, the term "magazine" also includes fireworks shipping buildings for display fireworks.

[3] For fireworks storage magazines in use prior to (30 days from the date of publication of the final rule in the Federal Register), the distances in this table may be halved if properly barricaded between the magazine and potential receptor sites.

[4] This table does not apply to the storage of bulk salutes. Use table at § 555.218.

[T.D. ATF–293, 55 FR 3723, Feb. 5, 1990, as amended by T.D. ATF–400, 63 FR 45004, Aug. 24, 1998]

Questions and Answers (Revised 1/07)
18 U.S.C. Chapter 40 and 27 CFR Part 555

Introduction

The following list of Questions and Answers are intended to aid you in gaining a better understanding of:

18 U.S.C. Chapter 40—Importation, Manufacture, Distribution and Storage of Explosive Materials and the implementing regulations issued within:

27 CFR Part 555—Commerce in Explosives

This listing is not all-inclusive. However it contains some of the most frequently asked questions that ATF receives. These questions and answers are intended only as a general overview. To determine how the law and regulations apply to your specific circumstances, you must refer directly to the applicable law and regulation or contact your local ATF Office of Industry Operations. Also, please be aware that both the law and regulations are subject to change. Please contact your local ATF office for the most up-to-date information. You can also find the latest Federal explosives regulations on the ATF website: www.atf.gov/.

Unless otherwise stated, these Questions and Answers apply only to Federal law and regulations. States and local jurisdictions have, in many cases, enacted their own requirements relating to explosives. Check with appropriate State or local authorities for information on those requirements. Compliance with Federal law and regulations does not exempt any person from compliance with any State or local requirements.

A Table of Contents and a Subject Index have been included for your convenience. The Index is located at the end of the Questions and Answers section.

Table of Contents

General Questions

1. Who is affected by the Federal explosives law?

The law affects all persons who import, manufacture, deal in, purchase, use, store, or possess explosive materials. It also affects those who ship, transport or cause to be transported, or receive explosive materials. Also, see 18 U.S.C. 845 and 27 CFR 555.141 for exemptions.

2. What changes were brought about by the Safe Explosives Act?

Among other things, the Act mandated that all persons who wish to receive or transport explosive materials must first obtain a Federal explosives license or permit. In addition, the act imposed new restrictions on who may lawfully receive and possess explosive materials (See question 4). All Federal explosive licensees and permittees and their responsible persons and employees authorized to possess explosives are affected by the new requirements and background checks mandated by the Act.

3. Can I obtain general information from ATF on the Internet?

Yes. ATF maintains a website on the Internet at www.atf.gov/.

4. Does the law make some classes of persons ineligible to receive a Federal license to import, manufacture, or deal in explosive materials or to receive a Federal explosives permit?

Yes. A license or permit will not be issued to any person who:

(a) Is under indictment for, or who has been convicted in any court, of a crime punishable by imprisonment for a term exceeding one year;

(b) Is a fugitive from justice;

(c) Is an unlawful user of or addicted to any controlled substance (as defined in section 102 of the Controlled Substances Act (21 U.S.C. 802));

(d) Has been adjudicated a mental defective or who has been committed to a mental institution;

(e) Is an alien (with certain exceptions);

(f) Has been discharged from the armed forces under dishonorable conditions; or,

(g) Having been a citizen of the United States, has renounced his citizenship. [18 U.S.C. 842(i), 843(b)(1); 27 CFR 555.49(b)(2)(i)]

5. Are there classes of persons to whom the distribution of explosive materials by licensees is prohibited?

Yes. No person shall knowingly distribute explosive materials to any individual listed in Question 4, or to an individual who is under 21 years of age. [18 U.S.C. 842(d); 27 CFR 555.26(d)]

6. What other distributions of explosive materials by licensees and permittees are prohibited?

A licensee or permittee shall not knowingly distribute any explosive materials to any person who:

(a) Is not a licensee [18 U.S.C. 842(b), 27 CFR 555.105, 555.106];

(b) Is not a holder of a user permit [18 U.S.C. 842(b); 27 CFR 555.105, 555.106];

(c) Is not a holder of a limited permit who resides in the same State where distribution is made and in which premises of the transferor are located. [18 U.S.C. 842(b); 27 CFR 555.105, 555.106];

(d) The licensee has reason to believe intends to transport such explosive materials into a State where the purchase, possession, or use of explosive materials is prohibited or which does not permit its residents to transport or ship explosive materials into the State or to receive explosive materials in the State. [18 U.S.C. 842(c); 27 CFR 555.106]

(e) Is in any State where the purchase, possession, or use by such person of such explosive materials would be in violation of any State law or any published ordinance applicable at the place of distribution. [18 U.S.C. 842(e); 27 CFR 555.106(b)(2)]

7. Does Federal law prohibit certain persons from receiving or possessing explosive materials?

Yes. The law prohibits the receipt or possession of explosive materials by any person listed in question 4. [18 U.S.C. 842(i); 27 CFR 555.26, 555.49(b)]

8. May a licensed manufacturer, importer, or dealer distribute explosive materials to nonlicensees and nonpermittees?

No. Every person who receives explosive materials must first obtain a Federal explosives license or permit. Distribution of explosive materials to persons who do not hold a license or permit is unlawful. Also, see 18 U.S.C. 845 and 27 CFR 555.141 for exemptions. [18 U.S.C. 842(a), 842(b); 27 CFR 555.26(a), 555.106]

9. Does Federal law provide penalties for purchasers who give false information at the time of purchasing explosive materials?

Yes. The penalty for knowingly providing false information or misrepresented identification is a maximum 10 years' imprisonment and/or a fine not exceeding $250,000. [18 U.S.C. 842(a)(2), 844(a)]

10. Is the theft of explosive materials, as well as the possession of stolen explosive materials, a Federal crime?

Yes. It is a Federal crime for any person to steal any explosive materials. It is also a Federal crime for any person to receive, possess, transport, ship, conceal, store, barter, sell, dispose of, or pledge or accept as security for a loan any stolen explosive materials. [18 U.S.C. 842(h), 844(k), (l)]

11. Are thefts and losses of explosive materials required to be reported to ATF?

Yes. Any licensee or permittee who has knowledge of the theft or loss of any explosive materials from his or her stock shall, within 24 hours of discovery, report the theft or loss by telephoning 800-461-8841 (Monday–Friday 8:00 a.m.–5:00 p.m. Eastern Time) or 888-283-2662 (after hours and weekends) (nationwide toll free number) and on ATF Form 5400.5, "Report of Theft or Loss—Explosive Materials", in accordance with the instructions on the form. The theft or loss shall also be reported to appropriate local authorities. The same requirements are imposed upon persons other than licensees and permittees, except that nonlicensees and nonpermittees, other than carriers, need not report a theft or loss on Form 5400.5, but must report the theft or loss by telephone, using the same numbers: 800-461-8841 (Monday–Friday 8:00 a.m.–5:00 p.m. Eastern Time) or 888-283-2662 (after hours and weekends) and in writing to the nearest ATF office. The theft or loss shall also be reported to appropriate local authorities. Carriers of explosive materials must report a theft or loss by telephone but need not make the report on the ATF form or in writing. See 27 CFR 555.30 for the specific information required to be reported in connection with a theft or loss. [18 U.S.C. 842(k), 18 U.S.C. 844(p); 27 CFR 555.30]

12. My company holds a Federal explosives license and after conducting an inventory of our explosives on hand, we noticed one case of dynamite missing. After double-checking all Daily Summaries of Magazine Transactions, invoices, and delivery sheets, we still cannot reconcile the discrepancy. What are we required to do?

This should be considered a theft or loss of explosive materials. As stated in the answer to Question 11, you must report the theft or loss of explosive materials, within 24 hours of discovery, to ATF by telephone (toll free: 800-461-8841 (Monday–Friday 8:00 a.m.–5:00 p.m. Eastern Time) or 888-283-2662 (after hours and weekends). ATF Form 5400.5, "Report of Theft or Loss—Explosive Materials", must then be completed and forwarded in accordance with the instructions on the form. [18 U.S.C. 842(k); 27 CFR 555.30]

13. May ATF conduct warrantless inspections of licensees' and permittees' records of explosives materials, stocks of such materials, and magazines?

Except for limited permit holders, any ATF officer may, without a warrant, enter during business hours the premises, including places of storage, of any licensee or permittee for the purpose of inspecting or examining any records or documents required to be kept by the law and regulations and any explosive materials kept or stored at the premises. For inspection purposes, "business hours" includes hours during which business is actually conducted, not just those hours stated on license applications. Any licensee or permittee who refuses to permit the inspection or examination is subject to having his or her license or permit revoked, as well as to denial of an application to renew the license or permit. For limited permit holders, an ATF officer may inspect the places of storage for explosive materials of either an applicant for a limited permit or at the time of renewal of such permit, but in no event shall such inspection occur more than once every three years. [18 U.S.C. 843(b)(4), 18 U.S.C. 843(f)]

14. Will ATF investigate accidents involving explosive materials?

ATF is authorized to inspect the site of any accident or fire where there is reason to believe that explosive materials were involved. Other Federal agencies, or State or local agencies, may also investigate such incidents, depending on the circumstances. [18 U.S.C. 846(a); 27 CFR 555.31]

15. Is black powder subject to regulation under Federal explosives laws?

Black powder is an explosive material for purposes of Federal explosives laws and regulations. However, the law exempts from regulation commercially manufactured black powder in quantities not exceeding 50 pounds (as well as percussion caps, safety and pyrotechnic fuses, quills, quick and slow matches, and friction primers) intended to be used solely for sporting, recreational, or cultural purposes in antique firearms as defined in 18 U.S.C. 921(a)(16) or in antique devices exempted from the term "destructive device" in 18 U.S.C. 921(a)(4). However, persons engaged in the business of importing, manufacturing, or dealing in black powder in any quantity must have a Federal explosives license. [18 U.S.C. 841(c), 841(d), 845(a)(5); 27 CFR 555.11: definitions of "explosives" and "explosive materials", 555.141(b)]

16. Is small arms ammunition subject to regulation under Federal explosives laws?

No. The law specifically exempts small arms ammunition and components thereof. (See also Question 81.) [18 U.S.C. 845(a)(4)]

17. Are binary explosives subject to regulation under Federal explosives laws?

Until the compounds are mixed, they are not classified as explosives and, therefore, are not subject to control. However, once mixed, binary explosives are "explosive materials" and are subject to all applicable Federal requirements. A person who mixes or combines compounds of binary explosives for the purpose of sale or distribution or for the person's own business use is a "manufacturer" of explosive materials and must be licensed as a manufacturer under the law. [18 U.S.C. 841(h); 27 CFR 555.11: definition of "manufacturer"]

18. Does ATF have any regulations governing the actual transportation of explosive materials?

Federal explosives laws and regulations generally prohibit any person from transporting explosive materials interstate or intrastate unless the person has a Federal explosives license or permit. Also, the transportation of stolen explosives materials is a Federal crime (see also Question 10). However, the law exempts from regulation under 18 U.S.C. Chapter 40 and 27 CFR Part 555 aspects of the transportation of explosive materials via railroad, water, highway, or air which are regulated by the United States Department of Transportation, and the Department of Homeland Security, and agencies thereof and which pertain to safety and security. [18 U.S.C. 842(a)(3), 842(h), 845(a)(1); 27 CFR 555.26, 555.28, 555.141(a)(1)]

19. Are common or contract carriers required to obtain a Federal explosives license or permit to transport explosive materials?

No. The actual transportation of explosive materials by carriers is subject to Department of Transportation or Department of Homeland Security regulations. [18 U.S.C. 845(a)(1); 27 CFR 555.141(a)(1)]

20. What is the "Explosives List"?

The Explosives List is a comprehensive (but not all-inclusive) listing of explosive materials which have been determined to be within the coverage of Chapter 40. The list is published annually by ATF (the most recent list can be found under the Arson & Explosives Publications link on the ATF website). [18 U.S.C. 841(d); 27 CFR 555.23]

21. May a person under the age of 21 be lawfully employed by an explosives business and lawfully receive, possess, and use explosive materials on behalf of the business?

Yes. Federal explosives law prohibits any person from distributing explosive materials to persons under 21 years of age. However, it does not prohibit the delivery to or possession of explosive materials by persons under the age of 21 who are receiving or using the materials on behalf of their employers to whom the materials were lawfully sold. [18 U.S.C. 842(d), (i); 27 CFR 555.11 (definition of "distribute"), 555.26, 555.106(b)(1)]

22. ATF regulations require explosive materials to be stored at certain minimum distances from a "public highway". What is a "public highway" for purposes of the regulations?

The term "highway" is defined in 27 CFR 555.11 as "any public street, public alley, or public road, including a privately financed, constructed, or maintained road that is regularly and openly traveled by the general public." Privately financed, constructed, or maintained roads that are marked and barricaded in a manner that prevents access by the general public do not fall within the meaning of the term and would, therefore, be exempt from table of distance requirements. [27 CFR 555.11: definition of "highway"] (See also ATF Ruling 2005–2)

23. Is an airport runway or taxiway considered a public highway for purposes of the Table of Distances for storage of explosive materials?

No. However, airport terminals are considered inhabited buildings for Table of Distance requirements.

24. How is shock tube regulated by ATF?

Shock tube contains highly explosive material. However, it may be stored as a low explosive when not attached to a detonator. [27 CFR 555.202(b), 555.213]

25. What is an EX number?

An EX number is a number, preceded by the prefix "EX–", which is issued and used by the Department of Transportation (DOT) to identify an explosive which has been tested and classified by DOT. See U.S. Department of Transportation regulations at 49 CFR 171.8 and 49 CFR 173.56.

26. What is a UN number?

A UN (United Nations) number is used by DOT as a method of identification and classification of products for shipping purposes. UN numbers are different from the hazard class or division designations used by DOT (for example, 1.1, 1.2, 1.3, 1.4, and 1.5). ATF regulations in 27 CFR Part 555 also use UN numbers to help identify certain explosives. [27 CFR 555.11]

27. Can Federal explosives disabilities resulting from a conviction of a crime punishable by imprisonment for a term exceeding one year be removed if the conviction is expunged or set aside or the convicted person has received a pardon for the offense or has had his or her civil rights restored?

A person convicted of, or under indictment for, a "crime punishable by imprisonment for a term exceeding one year" may not lawfully receive or possess explosive materials or be issued a Federal explosives license or permit. The term "crime punishable by imprisonment for a term exceeding one year" does not include offenses pertaining to antitrust violations, unfair trade practices, restraints of trade, or any State offense (other than one involving a firearm or explosive) classified as a misdemeanor and punishable by imprisonment for 2 years or less. There are only 3 means by which Federal explosives disabilities resulting from a conviction of, or indictment for, a "crime punishable by imprisonment for a term exceeding one year" can be removed:

(a) A decision of a court invalidating a conviction on the basis that the conviction was unconstitutional;

(b) In the case of a Federal conviction, a presidential pardon; and

(c) The granting of relief from Federal explosives disabilities by ATF pursuant to the filing of a relief application with the Director. Information on how to apply for relief and contact information for the Relief of Disabilities Section is available on the ATF website. [18 U.S.C. 841(l), 842(d),(i), 845(b); 27 CFR 555.11: definition of "crime punishable by imprisonment for a term exceeding one year", 555.26(c), 555.142]

Licenses and Permits

28. Who needs a Federal explosives license or permit?

All persons who wish to transport, ship, cause to be transported, or receive explosive materials must first obtain a Federal explosives license or permit. Certain exemptions apply. [18 U.S.C. 842(b); 18 U.S.C. 845; 27 CFR 555.26(a), 27 CFR 555.141]

29. Who is eligible for a Federal explosives license or permit?

The Chief, Federal Explosives Licensing Center, will approve a properly completed application for a license or permit on ATF Form 5400.13/5400.16 if the applicant:

(a) Is not a person prohibited from possessing or receiving explosive materials under 18 U.S.C. 842(i) and none of the applicant's "responsible persons" are prohibited under section 842(1); (see also Question 4 in General Q&A);

(b) Has not willfully violated any provision of Chapter 40 or the regulations in 27 CFR Part 555;

(c) Has not knowingly withheld information or has not made any false or fictitious statement intended or likely to deceive, in connection with the application,

(d) Has premises in a State from which he intends to conduct business or operations;

(e) Has storage for the class (as described in 27 CFR 555.202) of explosive materials described on the application;

(f) Has certified in writing that he is familiar with and understands all published State laws and local ordinances relating to explosive materials for the location in which he intends to do business;

(g) Has submitted the certificate required by section 21 of the Federal Water Pollution Control Act, as amended (33 U.S.C. 1341) [18 U.S.C. 843(b); 27 CFR 555.49(b)];

(h) None of the applicant's employees authorized to possess explosives are prohibited persons under 18 U.S.C. 842(i); and

(i) In the case of an applicant for a limited permit, the applicant has certified that the applicant will not receive explosive materials on more than 6 occasions during the 12-month period for which the limited permit is valid.

30. What activities are covered by licenses and permits?

Licenses allow persons to engage in the business of importing, manufacturing, or dealing in explosive materials. Any individual or business entity intending to engage in any of these activities must first obtain a license. A user permit allows the receipt and transportation of explosive materials. A limited permit allows the receipt of explosive materials from a licensee or permittee within the permittee's state of residence only, and on no more than six occasions in the 12-month period during which the limited permit is valid. A limited permit does not authorize the receipt or transportation of explosive materials in interstate or foreign commerce. [27 CFR 555.11: definitions of "importer", "manufacturer", "dealer", "limited permit", and "user permit", 555.41] A separate license is needed for each business premises where an explosives business or activity is conducted. Only a single user permit is needed by a permittee who uses explosives in more than one location. [27 CFR 555.41]

31. What is a Limited Permit?

A limited permit is for persons who wish to transport, ship, cause to be transported, or receive explosive materials in intrastate commerce only. This permit is designed for the infrequent receipt of explosive materials by intrastate users. The limited permit will allow a purchaser to receive explosive materials on no more than six separate occasions from in-state licensees or permittees during the 12-month period of the permit. The limited permit does not allow the holder to transport, ship, cause to be transported, or receive explosive materials in interstate commerce.

32. What is the duration of a license or permit?

(a) A user license or permit is valid for a period of 3 years.

(b) The user-limited permit is valid only for a single purchase transaction.

(c) Limited permits are valid for no more than six separate receipts of explosive materials during a 12-month period. [27 CFR 555.51]

33. What are the fees for licenses and permits?

Each license applicant must pay a fee of $200 for obtaining a 3-year license, a separate license and fee being required for each business premises. The fee for renewal of a license is $100 for a 3-year license. [27 CFR 555.42]

Each applicant for a user permit must pay a fee of $100 for a 3-year permit, and each applicant for a user-limited permit (nonrenewable) must pay a fee of $75. The fee for renewal of a user permit is $50 for a 3-year permit. [27 CFR 555.43] Each applicant for a limited permit must pay a fee of $25 for a 1-year limited permit. The fee for renewal of a limited permit is $12 for a 1-year limited permit. [27 CFR 555.43]

34. Will the Government investigate an application for a license or permit?

ATF will investigate any applicant before issuing a license or permit. Additionally, ATF must inspect places of storage and conduct background checks on responsible persons and employee possessors authorized to possess explosives [18 U.S.C. 843(b); 18 U.S.C. 843(h); 27 CFR 555.33, 27 CFR 555.49(b)]

35. What may a licensed explosives dealer do?

A licensed dealer may engage in the business of distributing explosive materials at wholesale or retail [27 CFR 555.11: definition of "dealer"]

36. What may a licensed explosives importer do?

A licensed importer may engage in the business of importing or bringing explosive materials into the United States for purposes of sale or distribution. It is not necessary for a licensed importer to also obtain a dealer's license to engage in business on his or her licensed premises as a dealer in explosive materials (see also Question 52 and 53) [27 CFR 555.11: definition of "importer", 27 CFR 555.41(b)(2)]

37. When is a manufacturer's license required?

A manufacturer's license is required by persons engaged in the business of manufacturing explosive materials for sale, distribution, or for their own business use. For example, persons engaged in the business of providing a blasting service using explosives of their own manufacture would be required to have a manufacturer's license. Persons who manufacture explosives for their personal, non-business use are not required to have a manufacturer's license. However, no person may ship, transport, cause to be transported, or receive explosive materials unless such person holds a license or permit. [27 CFR 555.11: definition of "manufacturer", 555.41(b)] A separate manufacturer's license is not required by a licensed manufacturer for the purpose of on-site manufacture, for example, mixing binary explosives or making blasting agents at a quarry or other job site. It is not necessary for a licensed manufacturer to also obtain a dealer's license to engage in business on his or her licensed premises as a dealer in explosive materials (see also Question 52 and 53) [27 CFR 555.11: definition of "manufacturer", 555.41(b)(2)]

38. How do I apply for a Federal explosives license or permit?

You can request an application for a Federal explosives license or permit from the Federal Explosives Licensing Center at 877-283-3352 or from the ATF Distribution Center at 202-648-6420. As part of the application process, you must complete and submit an ATF Form 5400.13/5400.16, Application for Explosives License or Permit. You must also submit the names, identifying information, fingerprints, and photographs of all responsible persons. In addition, you must submit the names and identifying information of all employees who are authorized to possess explosive materials in the course of their employment on ATF Form 5400.28, Employee Possessor Questionnaire. [27 CFR 555.45(c)]

39. Who is a "responsible person"?

Federal explosives laws define a "responsible person" as an individual who has the power to direct the management and policies of the applicant pertaining to explosive materials. Responsible persons generally include sole proprietors and explosives facility site managers. In the case of a corporation, association, or similar organization, responsible persons generally include only those corporate directors/officers, and stockholders, who have the power to direct management and policies as they pertain to explosive materials.

For example, a corporate vice president whose duties include acquiring and approving contracts with explosives distributors would be considered a responsible person. Other corporate officials whose duties do not include the power to direct the management and policies of the applicant pertaining to explosive materials, for example, a vice president responsible solely for human resources, would not typically be considered a responsible person. Each applicant for a license or permit must assess the corporate and other management responsibilities for all key personnel and determine whether or not these duties place the individual in the position of being a responsible person. [18 U.S.C. 841(s), 27 CFR 555.11: definition of "responsible person"]

40. Who is a "possessor of explosives"?

A possessor of explosives is any employee of a license or permit holder or any employee of an applicant for a license or permit who has or will have actual physical possession of explosive materials or who has or will have constructive possession of explosive materials. For example, persons who physically handle explosive materials would be considered to be actual possessors of explosive materials. This would include employees who directly handle explosive materials as part of the production process; employees who handle explosive materials in order to ship, transport, or sell them; and employees, such as blasters and their helpers who actually use explosive materials. A constructive possessor is any person who has access to explosive materials, without physically handling them. For example, a supervisor at a construction site who keeps keys for storage magazines in which explosives are stored or who directs the use of explosive materials by other employees has constructive possession of explosives.

41. Why is it necessary to provide new and additional information on responsible persons and employee possessors of explosives?

The law requires this information for ATF to conduct background checks on all responsible persons and employee possessors to restrict the availability of explosives to authorized persons only and to reduce the risk of prohibited persons acquiring explosive materials. [18 U.S.C. 843(h); 27 CFR 555.33, 555.45(c)]

42. When will I need to submit the identifying information for my responsible persons and employee possessors of explosives?

(a) All license and permit applicants and any renewal applicants must submit identifying information for responsible persons and employee possessors (and fingerprints and photographs for responsible persons) upon submission of an original or renewal application.

(b) Any new responsible person added after a license or permit has been issued by ATF must be reported to ATF within 30 days. However, the submission of fingerprints and photographs by the new responsible person is required only at the time of any subsequent renewal.

(c) For all licenses and permits (new and renewal), any new employee possessors must be reported to ATF within 30 days of hire on the Employee Possessor Questionnaire form (ATF F 5400.28). [27 CFR 555.45(c), 27 CFR 555.57(b)]

43. How do I get my fingerprints taken?

Fingerprints must be submitted on Fingerprint Identification Cards, FD–258 that have been issued by ATF. The fingerprint cards must contain the following ORI information: WVATF0900; ATF–NATL EXPL LIC, MARTINSBURG WV. These fingerprint cards may be obtained by contacting the Federal Explosives Licensing Center at 877-283-3352 or the ATF Distribution Center at 202-648-6420. The fingerprint cards must be completed by your local law enforcement authority.

44. Will ATF notify me whether or not my responsible persons and employee possessors have passed their background checks?

Yes. A "Notification of Clearance" will be issued directly to all license or permit holders advising whether their responsible persons and employee possessors have been cleared to possess explosive materials, or are or may be prohibited from possessing

explosives. These notices must be retained as part of the license or permit holders permanent records. In addition, letters of clearance or denial will be issued directly to responsible persons and employee possessors. [27 CFR 555.33]

45. What notification will I receive if one of my responsible persons or employee possessors does not pass their ATF background check?

If an individual does not pass the background check, a letter will be sent to the licensee or permittee who submitted the individual's name indicating that the individual was denied. A letter will also be sent to that individual explaining the prohibition and outlining appeal and relief procedures, as may be applicable. Unless and until an appeal overturns the denial or relief from disabilities is granted, that individual may not lawfully possess explosives. [27 CFR 555.33]

46. Who will conduct the background checks on applicants, responsible persons, and possessors?

ATF will perform the background checks. If employers wish to require their own background checks as a condition of employment, they may do so. However, such a background check will not be accepted in place of the ATF background check. [27 CFR 555.33]

47. May I sell black powder without a license?

No. Anyone who engages in the business of selling black powder, regardless of quantity, must be licensed as an explosives dealer. [27 CFR 555.41(b)]

48. Is a manufacturer's license required to acquire and mix binary explosives?

If the individual purchasing the binary explosives is engaged in the business of manufacturing explosives, i.e., mixes and uses them in the operation of a commercial business (for example, operating a quarry, or providing the service of removing stumps or boulders from a farm field), then a manufacturer's license is required.

An individual farmer who merely wishes to mix the binary explosives to remove obstacles from his field and provides no other outside service would not need a manufacturer's license.

Please note, however: A Federal explosives license or permit would be required to obtain any explosive device, such as detonators, used to initiate the mixed binary explosives. In addition, transportation of any explosive material, including mixed binary explosives, without a Federal license or permit is prohibited. [27 CFR 555.11: Definition of "manufacturer"; 27 CFR 555.26, 555.41(b)]

49. What is theatrical flash powder and is there a license for its manufacture?

Theatrical flash powder is flash powder commercially manufactured in premeasured kits not exceeding 1 ounce in weight, and mixed immediately prior to use and intended for use in events such as theatrical shows, stage plays, band concerts, magic acts, thrill shows, and clown acts in circuses. A manufacturer's license allows on-site manufacturers to operate nationally on one license issued to their principal place of business. [27 CFR 555.11: definitions of "flash powder" and "theatrical flash powder", 555.41(b)]

50. Is a separate license required for each location where business is conducted?

Yes. A separate license is required for each location where business is conducted. However, a separate license is not required for:

(a) Facilities used only for the storage of explosive materials;

(b) Locations used solely for the storage of records relating to the business; and

(c) Licensed manufacturers' on-site manufacturing. [27 CFR 555.41(b)]

51. Must a person who engages in the business of both manufacturing and importing at the same location have both licenses?

Yes. The licenses for manufacturing and importing allow a person to engage in separate and distinct activities and a separate license is required for each activity. However, a manufacturer or an importer does not need a separate dealer's license to also distribute explosive materials from the licensed premises. [27 CFR 555.41(b)]

52. Does a licensed manufacturer, importer, or dealer need a permit to use explosive materials?

No. No licensee will be required to obtain a user permit to lawfully transport, ship, or receive explosive materials in inter-state or foreign commerce. [27 CFR 555.41(b)(2)]

53. Does a Federal license or permit exempt the holder from State or local requirements?

No. A license or permit confers no right or privilege to conduct business or operations, including storage, contrary to State or other law. All legal requirements must be followed, whether Federal, State, or local. [18 U.S.C. 848; 27 CFR 555.62]

54. Who is authorized to import explosive materials?

Any licensed importer is authorized to engage in the business of importing explosive materials for sale, distribution, or their own use. Any licensed manufacturer, dealer, or holder of a user permit may import explosive materials for their own use only. Licensees and user permittees importing explosive materials must provide to the U.S. Customs and Border Protection (CBP) a copy of the license or permit. Note, however, that in the case of certain military explosives or propellant powder or other components of

small arms ammunition, Federal firearms regulations require the importer to provide an approved ATF Form 6 to the CBP. [27 CFR 555.41(b)(2), 555.41(b)(3), 447.21, 555.108(a), 555.183, 478.113]

55. How may an employee of an explosives licensee or permittee qualify to accept delivery of explosive materials for the employer?

The employee must be on the current list of representatives or agents authorized to accept delivery of explosive materials on behalf of the employer and be an authorized employee possessor of explosives. [27 CFR 555.103(b), 555.105(b)]

56. When an explosives licensee or permittee sends one of their truck drivers to the distributor's premises to pick up explosive materials that have been purchased by the licensee or permittee, will the driver be required to sign any forms?

No, however the driver is required to furnish the seller with an identification document as defined in 27 CFR Part 555.11. [27 CFR 555.103(b), 555.105(b)]

57. Will a licensee or permittee be notified in advance when the license or user permit needs to be renewed?

Generally, prior to expiration of the license or permit, a licensee or permittee will be notified. The application form must be completed and filed with ATF before expiration of the current license or permit for the renewal to be considered timely. However, if a licensee or permittee does not receive a renewal notification, it is still that licensee's or permittee's responsibility to ensure that an application is filed prior to expiration of the current license or permit. [27 CFR 555.46]

58. I have timely filed my application for renewal of my license (or user's permit) but I have not received my new license (or permit). May I continue in business even though the expiration date shown on my license or permit has passed? If so, for how long?

Yes. You may continue to operate the business pursuant to your current license or permit until the application for renewal is acted upon. [5 U.S.C. 558]

59. Can a license or permit be revoked?

Yes. The Director, Industry Operations for the ATF Field Division in which a licensee or permittee is located may revoke a license or permit if the holder has violated any provision of 18 U.S.C. Chapter 40 or its implementing regulations or has become ineligible to receive explosive materials under 18 U.S.C. 842(i). [18 U.S.C. 843(d); 27 CFR 555.71, 555.74]

60. If a Federal explosives licensee or permittee is indicted for or convicted of a "crime punishable by imprisonment for a term exceeding one year", may he or she continue operations under the license or permit?

As stated in the answer to Question 4 in General Q&A, a person under indictment for, or convicted of, a crime punishable by imprisonment for a term exceeding one year is not eligible to be issued a license or permit. However, a licensee or permittee who is indicted for, or convicted of, such a crime during the term of his or her existing license or permit is not barred from licensed or permit operations for 30 days after the date of the indictment or the date the conviction becomes final. If the licensee or permittee files an application for relief from disabilities within such 30-day period, he or she may continue licensed or permit operations while the application is pending. If a relief application is not filed during that period, the licensee or permittee may not continue operations beyond such 30-day period. The right of a licensee to continue licensed or permitted operations beyond such 30-day period is also conditioned on the licensee or permittee timely filing a license or permit renewal application disclosing that the applicant has been indicted for, or convicted of, the crime. A licensee or permittee may not continue operations beyond 30 days following the date the Director issues notification that the relief application has been denied. [18 U.S.C. 845(b); 27 CFR 555.142]

61. May a licensed dealer make a sale to a holder of a limited permit in an adjoining State?

No. Sales may not be made to limited permittees who are out-of-State residents. [18 U.S.C. 842(a); 27 CFR 555.11: definition of "limited permit", 555.41(b)(3)]

Recordkeeping

62. Does a licensee or permittee have to keep records of the acquisition, distribution, and storage of explosive materials?

Yes. Licensees and permittees must keep records of acquisitions, dispositions, and storage of explosive materials. [18 U.S.C. 842(f), 847; 27 CFR 555.107, 555.122–.125, and 555.127, Subpart G]

63. How do licensees and permittees account for explosive quantities in their records?

If acquisitions are recorded by weight, then distribution must also be recorded by weight. If acquisitions are recorded by physical count (e.g., by units), then distribution must also be recorded by physical count. [27 CFR 555.122–.125]

64. Must a licensee or permittee maintain a daily summary of magazine transactions?

Yes. After the initial inventory required by regulations has been taken, the inventory shall be entered in a record of daily transactions. Not later than the close of the next business day, each licensee and permittee shall record by manufacturer's name or brand name the total quantity received in and removed from each magazine during the day and the total remaining on hand at the end of the day. [27 CFR 555.127]

65. Where must a licensee or permittee keep the daily summary of magazine transactions?

The records must either be kept at each magazine or at one central location on the business premises, provided a separate record of daily transactions is maintained for each magazine. [27 CFR 555.127]

66. How can I obtain additional copies of ATF Forms?

Forms are available on-line at http://www.atf.gov/forms/explosives/. Requests for forms should be mailed to the ATF Distribution Center, 1519 Cabin Branch Drive, Landover, MD 20785. You may also have forms mailed to you by submitting an on-line request at http://www.atf.gov/forms/dcof/, or by telephoning your request to (202) 648-6420. [27 CFR 555.21(b)]

67. Does a purchaser of black powder have to sign any forms at the time of purchase?

If 50 pounds or less of commercially manufactured black powder is being purchased, and the powder is intended to be used solely for sporting, recreational, or cultural purposes in antique firearms as defined in 18 U.S.C. 921(a)(16) or in antique devices exempt from the term "destructive device" in 18 U.S.C. 921(a)(4), no form is required. However, if the black powder is being purchased for any other purpose (regardless of quantity), the purchaser or other transferee must possess a Federal explosives license or permit. [18 U.S.C. 845(a)(5); 18 U.S.C. 926(c); 27 CFR 555.141(b), 555.26(a)]

68. Is there a requirement for licensees and permittees to make an annual inventory of explosive materials on hand?

Yes. An inventory is required to be taken at least once a year. [27 CFR 555.122–.125]

69. When must ATF Form 5400.4, "Limited Permittee Transaction Report (LPTR)", be executed?

Before distribution of explosive materials to a limited permittee, the licensee or permittee must obtain an executed ATF F 5400.4 from the limited permittee with an original unaltered and unexpired Intrastate Purchase of Explosives Coupon (IPEC) attached. Except when delivery of explosive materials is made by a common or contract carrier who is an agent of the limited permittee, the licensee or permittee must verify the identity of the holder of the limited permit by examining an identification document (as defined in 555.11) and noting on the ATF F 5400.4 the type of document presented. The licensee or permittee must complete the appropriate section on ATF F 5400.4 to indicate the type and quantity of explosive materials distributed, the license or permit number of the seller, and the date of the transaction. The licensee or permittee must sign and date the form. [27 CFR 555.126(b)]

70. Do the ATF Forms 5400.4 have to be maintained by the licensee or permittee making the sale?

Yes. One copy of ATF F 5400.4 must be retained by the seller as part of his permanent records in chronological order by date of disposition, or in alphabetical order by name of limited permittee. They must be maintained for a period of five years. [27 CFR 555.126]

71. May I keep computerized records?

Yes. See ATF Ruling 2007–1.

Storage

72. Who must comply with the storage requirements?

Except for those items and activities given exempt status under 18 U.S.C. 845 (also see 27 CFR 555.141), or exempted under 27 CFR 555.32, Special Explosives Devices, all persons who store explosive materials must store them in conformity with the provisions of Subpart K of the regulations, unless the person or the materials are exempt from regulation. [18 U.S.C. 842(j); 27 CFR 555.29, 555.141, 555.201(a)]

73. What are the classes of explosive materials for storage purposes?

There are 3 classes of explosive materials:

(a) High explosives (for example, dynamite, flash powders, and bulk salutes);

(b) Low explosives (for example, black powder, safety fuses, igniters, igniter cords, fuse lighters, and "display fireworks", except for bulk salutes); and

(c) Blasting agents (for example, ammonium nitrate-fuel oil and certain water gels). [27 CFR 555.202]

74. May a person store explosive materials in a residence or dwelling?

No. Storage of explosive materials in a residence or dwelling is prohibited. [27 CFR 555.208(b), 555.210(b), 555.211(b)]

75. What is the "Table of Distances"?

This table lists the minimum acceptable distances separating explosives magazines from inhabited buildings, passenger railroads, public highways, and other explosives magazines. The table is contained in 27 CFR 555.218.

76. When low and high explosives are stored together, how is the distance determined to meet the table of distance requirements?

The table of distances for high explosives at 27 CFR 555.218 would be applied using the total weight of explosive materials in the magazine. [27 CFR 555.218]

77. Is it necessary to inspect my explosives magazines on a regular basis?

Yes. Any person storing explosives must inspect the magazines at least once every 7 days to determine whether there has been unauthorized entry or attempted entry into the magazines or unauthorized removal of the contents of the magazines. [27 CFR 555.204]

78. What are the requirements for making changes or additions to an approved storage facility?

Making changes in construction to an approved explosives magazine or adding a magazine requires that ATF be notified. However, mobile or portable type 5 magazines and magazines used for the temporary (under 24 hours) storage of explosive materials are exempt from this requirement. See 27 CFR 555.63 for details.

79. Is any type of black powder fuse exempt from storage requirements?

Yes, 3⁄32" and other external burning pyrotechnic hobby fuses are exempt from the requirements of Federal explosives laws and regulations. [18 U.S.C. 845(a)(4–5); 27 CFR 555.11: definition of "ammunition", 555.141(a)(4), 555.141(b)]

80. With the exception of 3⁄32" pyrotechnic safety fuse for use in small arms, must black powder fuses generally be stored in approved explosives magazines?

Yes. Generally igniter fuses, time fuses, blasting fuses, safety fuses, or other black powder fuses by whatever name known, must be stored in approved magazines.

81. Is smokeless powder designed for use in small arms ammunition subject to the explosives storage requirements?

Smokeless propellants designed for use in small arms ammunition are exempt from regulation under 18 U.S.C. Chapter 40 and the regulations in 27 CFR Part 555. However, it should be noted that persons engaged in the business of importing or manufacturing smokeless propellants must have a Federal explosives license. Additionally, smokeless propellant designed for use other than small arms ammunition is not exempt. Therefore, explosives products such as squibs, fireworks, theatrical special effects, or other articles that may be utilizing smokeless propellants are regulated and must be stored accordingly.

82. My office building, in which several company employees work during the day in connection with my explosives business, is located in the general area of my explosives magazine. Do the regulations and the Table of Distances apply to this building as an "inhabited building"?

No. A building such as an office building or repair shop which is part of the premises of an explosives business and is used by the business in connection with the manufacture, transportation, storage, or use of explosive materials is not considered to be an "inhabited building". [27 CFR 555.11: definition of "inhabited building", 555.218]

83. Am I required to notify my State or local authorities about my explosives storage magazines?

Yes. All persons who store explosive materials must notify the fire department having jurisdiction over the site where explosive materials are manufactured or stored. Notification must be made orally by the end of the day on which storage begins and in writing within 48 hours from the time storage began. The notification must include the type of explosive materials, magazine capacity, and the location of each storage site. [27 CFR 555.11: Definition of "authority having jurisdiction for fire safety", 27 CFR 555.201(f)]

84. What is the definition of a "case hardened shackle?"

Case hardening involves putting carbon (or a combination of carbon and nitrogen) into the surface of the steel to make it a high-carbon steel, which can be hardened by heat treatment. Only the outer skin gets hard in this manner. The center is still tough and malleable. This makes for a strong lock with a tough surface.

85. Can detonators be stored with detonating cord?

No. However, products which are manufactured with a detonator attached to the detonating cord as an integral part need not be disassembled and stored separately. [27 CFR 555.213]

86. Are there storage requirements for oxidizers, such as ammonium nitrate?

In general, no. However, when a magazine or bin containing ammonium nitrate is located within the sympathetic detonation distance of other explosives or blasting agents, it must be stored in accordance with the table of distances in 27 CFR 555.220.

87. Are State and local government agencies required to store their explosive materials in conformity with Federal storage regulations?

Yes. There is no exemption in the law or regulations for the storage of explosive materials by any State or political subdivision thereof. [18 U.S.C. 842(j), 845(a)(6); 27 CFR 555.141(a)(3), (a)(5)]

Fireworks

Fireworks are defined in the Federal explosives regulations as any composition or device designed to produce a visible or an audible effect by combustion, deflagration, or detonation. Fireworks are further divided into two broad classifications, consumer fireworks or display fireworks as defined at 27 CFR Part 555.11.

88. Are "consumer fireworks" subject to regulation under the Federal explosives laws?

No. The importation, distribution, and storage of fireworks defined as consumer fireworks are exempted from the provisions of the Federal explosives laws. However, because they contain pyrotechnic compositions classed by ATF as explosive materials, the manufacture of consumer fireworks requires a manufacturer's license. In addition, pyrotechnic compositions used in the manufacture of consumer fireworks must be stored in accordance with regulations in 27 CFR Subpart K. Consumer fireworks are defined as "any small firework device designed to produce visible effects by combustion and which must comply with the construction, chemical composition, and labeling regulations of the U.S. Consumer Product Safety Commission, as set forth in Title 16, Code of Federal Regulations, parts 1500 and 1507. Some small devices designed to produce audible effects are included, such as whistling devices, ground devices containing 50 mg or less of explosive materials, and aerial devices containing 130 mg or less of explosive materials. Consumer fireworks are classified as fireworks UN0336 and UN0337 by the U.S. Department of Transportation at 49 CFR 172.101. This term does not include fused set pieces containing components which together exceed 50 mg of salute powder." [27 CFR 555.11: definition of "consumer fireworks"; definition of "licensed manufacturer", 555.141(a)(7)]

89. Are "display fireworks" considered to be explosive materials subject to regulation under Federal explosives laws and regulations?

Yes. Display fireworks include, but are not limited to, salutes containing more than 2 grains (130 mg) of explosive materials, aerial shells containing more than 40 grams of pyrotechnic compositions (excluding the lift charge), and other display pieces which exceed the limits of explosive materials for classification as "consumer fireworks". These fireworks are classified as fireworks UN0333, UN0334, or UN0335 by regulations of the U. S. Department of Transportation at 49 CFR 172.101. Display fireworks also include fused set pieces containing components which together exceed 50 mg of salute powder. [27 CFR 555.11: definition of "display fireworks"]

90. How must display fireworks be stored?

Display fireworks, with the exception of bulk salutes, are considered low explosives and, at a minimum, must be stored in type 4 storage magazines. They may also be stored in type 1 or type 2 magazines. Bulk salutes, which are defined as either salute components prior to final assembly into aerial shells, (or) finished salute shells held separately prior to being packed with other types of display fireworks, are classified as high explosives. As such, bulk salutes may only be stored in type 1 or type 2 magazines specifically constructed for the storage of high explosives. [27 CFR 555.11, 555.202(b), 555.203(d), 555.207, 555.208, 555.210]

91. Are "Articles Pyrotechnic" subject to the requirements of the Federal explosives regulations?

The importation, distribution, and storage of fireworks defined as "Articles Pyrotechnic", are exempt from the Federal explosives laws and regulations. However, because they contain pyrotechnic compositions classed by ATF as explosive materials, the manufacture of items defined as "articles pyrotechnic" requires an ATF manufacturer's license. In addition, pyrotechnic compositions used in the manufacture of articles pyrotechnic must be stored in accordance with regulations in 27 CFR Subpart K. [27 CFR 555.11: definitions of "articles pyrotechnic" and "consumer fireworks", 555.141(a)(7)]

92. Must partially assembled display fireworks be removed from a drying building for overnight storage?

Yes. At the end of a day's manufacturing operations, all dry explosive powders and mixtures and partially assembled and finished display fireworks must be removed from fireworks process buildings and stored in a magazine meeting the storage requirements in 27 CFR Part 555, Subpart K. [27 CFR 555.205, 555.221]

93. What areas of a fireworks manufacturing plant are considered to be "fireworks process buildings?"

Fireworks process buildings include any buildings in which pyrotechnic compositions or explosives materials are mixed, pressed, finished, or assembled. Fireworks process buildings do not include plant warehouses, office buildings, or other buildings and areas in which no fireworks, pyrotechnic compositions, or explosive materials are processed or stored. [27 CFR 555.11: definition of "fireworks process building"]

94. Under what conditions may I temporarily store display fireworks (including low explosives for choreographed shows) on trucks?

See ATF Ruling 2007–2.

95. What types of fireworks require an ATF license or permit in order to be lawfully transported or received?

Any fireworks defined as "display fireworks" in 27 CFR 555.11 may be lawfully received or transported only by persons who hold a valid license or permit. No ATF license or permit is required to receive or transport "consumer fireworks" or "articles pyrotechnic". [18 U.S.C. 842(a)(3); 27 CFR 555.26, 555.141(a)(7)]

Plastic Explosives

96. What is a plastic explosive?

A plastic explosive is defined as "an explosive material in flexible or elastic sheet form formulated with one or more high explosives which in their pure form has a vapor pressure less than $10<4>$ Pa at a temperature of 25 °C, is formulated with a binder material, and is as a mixture malleable or flexible at normal room temperature." [18 U.S.C. 841(q); 27 CFR 555.180(d)(4)]

97. What plastic explosives are required to contain detection agents?

All plastic explosives manufactured or imported on or after April 24, 1996, must contain a detection agent. Federal law enforcement agencies and the military may possess unmarked plastic explosives if they meet the requirements of the use-up period described in Question 103. [18 U.S.C. 841(q), 842(n); 27 CFR 555.180]

98. What are the permissible detection agents for marking plastic explosives?

These agents are listed in the law and regulations at 18 U.S.C. 841(p) and 27 CFR 555.180(d)(3).

99. Is it lawful to manufacture plastic explosives that do not contain a detection agent?

No. [18 U.S.C. 842(l); 27 CFR 555.180(a)]

100. Is it lawful to import into the United States plastic explosives that do not contain a detection agent?

No. The importation of plastic explosives into the United States requires that the importer file ATF Form 6 certifying that the imported plastic explosives contain the required detection agent, or is exempted from the marking requirements as provided in the regulations. [18 U.S.C. 842(m); 27 CFR 555.180(b), 555.182, 555.183]

101. Is it lawful to ship, transport, transfer, receive, or possess any plastic explosive that does not contain a detection agent?

No. However, a 15-year use-up period is provided for Federal law enforcement agencies and the military for unmarked plastic explosives imported into or manufactured in the U.S. prior to April 24, 1996. Any stocks of unmarked plastic explosives in their possession must be used, destroyed or properly marked by June 21, 2013 [18 U.S.C. 842(n); 27 CFR 555.180(c)]

102. If a person acquired plastic explosives not containing a detection agent before April 24, 1996, may he or she continue to lawfully possess the explosives?

No. With the exception of the use-up period provided by law for Federal law enforcement agencies or the military, the time period for lawful possession of unmarked plastic explosives terminated on April 24, 1999. [18 U.S.C. 842(n); 27 CFR 555.180(c)]

103. Are police departments exempt from the prohibition against possessing unmarked plastic explosives after April 24, 1999?

No. Police departments and other State or local law enforcement agencies could lawfully possess unmarked plastic explosives acquired on or before April 24, 1996, until April 24, 1999. Such agencies still possessing unmarked plastic explosives should destroy them or abandon them to ATF. Contact the nearest ATF field office for information. [18 U.S.C. 842(n); 27 CFR 555.180(c)(1)]

U.S. Military Explosives

104. Would an ATF license or permit be needed to demilitarize (demil) U.S. military explosives?

As long as the demil operator has a valid Department of Defense contract to perform such operations, the operations would be exempt from 27 CFR Part 555 and no license or permit would be required. However, if title to the explosive materials has passed from the military to the demil operator and the operator intends to resell the explosives on the commercial market, then such operations may be regulated by ATF (e.g., storage, sales, manufacturing) and an ATF license or permit may be needed. Contact the nearest ATF field office for further information. [18 U.S.C. 845(a)(3), (a)(6) and 27 CFR 555.141(a)(3), (a)(5)]

105. Would a civilian contractor who is manufacturing explosive materials pursuant to a government contract for or on behalf of the United States military be entitled to the exemptions from the explosives laws and regulations?

Yes, provided that all the explosive materials in question are manufactured under a government contract. Any explosive materials manufactured in anticipation of receiving a government contract would not qualify for this exemption.

If the contractor manufactures any explosive materials not pursuant to a U.S. military contract, the manufacture and the explosive materials are subject to all requirements of the law and regulations. [18 U.S.C. 845(a)(3), (a)(6); 27 CFR 555.26, 555.41, 555.141(b)]

106. Is an ATF licensee or permittee, whose licensed premises are located on a U.S. military installation, subject to the regulations in 27 CFR Part 555?

All activities conducted outside the scope of a U.S. Government contract are subject to the requirements of Part 555, even if the activities are conducted on property owned by the military. [18 U.S.C. 845(a)(3), (a)(6); 27 CFR 555.26, 555.29, 555.41, 555.141(a)(3), (a)(5)]

Index to Questions and Answers

ATF Explosives Rulings (Revised 3/12)

Table of Contents

1. 27 CFR 181.11: Meaning of Terms (Also 181.186)

An office or repair shop used in connection with the manufacture, etc., of explosive materials are not "inhabited buildings."

ATF Ruling 75–20

The Bureau of Alcohol, Tobacco and Firearms has been asked to explain the application of 27 CFR 181.11, as it relates to "Inhabited Building."

The case in question concerns whether or not a building, such as an office or repair shop, which is located on manufacturing premises closer to facilities approved for the storage of explosive materials than permitted by the American Table of Distances, as set forth in regulations implementing Title II, Regulation of Explosives (Chapter 40, Title 18, U.S.C.), is an "inhabited building," as defined in 27 CFR 181.11)

Regulations in 27 CFR 181.186 and 181.198 set forth provisions concerning the location of storage facilities and the minimum distances such storage facilities may be located from, among other things, "inhabited buildings."

Regulations in 27 CFR 181.11 define "inhabited building" as any building regularly occupied in whole or in part as a habitation for human beings, or any church, schoolhouse, railroad station, store or other structure where people are accustomed to assemble, except any building occupied in connection with the manufacture, transportation, storage, or use of explosive materials.

These provisions are intended to provide protection to persons who inhabit buildings located near premises where explosives are manufactured, stored, etc. However, it is the intent of section 181.11 to except buildings used by the explosives industry in connection with the manufacture, transportation, storage, or use of explosive materials from the table of distance requirements on "inhabited buildings."

Held, a building, such as an office or repair shop, which is a part of the premises of an explosives manufacturer and is used in connection with the manufacture, transportation, storage, or use of explosive materials is not an "inhabited building" as defined in 27 CFR 181.11.

Signed: June 26, 1975

[Editor's Note: 27 CFR Part 181 is now 27 CFR Part 555.]

2. 27 CFR 181.187: Construction of Type 1 Storage Facilities (Also 181.190)

Certain explosives storage facilities meeting standards of construction prescribed by the Department of Defense Explosives Safety Board for such storage are approved by the Bureau.

ATF Ruling 75–21

The Bureau of Alcohol, Tobacco and Firearms has been asked to state its position with respect to concrete floors used in certain types of explosives storage facilities. Specifically, the question has been raised whether concrete floors of Type 1 storage facilities manufactured for the Department of Defense and currently being leased to licensees and permittees for the storage of commercial explosives may be considered nonsparking under the provisions of 27 CFR Part 181.

Regulations in 27 CFR 181.187 and 181.190, which implement, in part, title II, Regulation of Explosives (18 U.S.C. Chapter 40), provide in pertinent part that floors of Type 1 and Type 4 storage facilities for the storage of explosives shall be constructed of or covered with a nonsparking material.

The Bureau has been advised by the Department of Defense Explosives Safety Board that the majority of explosives magazines constructed for the Department of Defense have smooth finished concrete floors and that current Defense Department specifications for such magazines have no requirement for sparkproof flooring. Although these magazines are approved by the Board for the storage of all types of explosive materials, the Board does recognize that the Department of the Navy advises that black powder be stored in explosives magazines with spark resistant floor finishes.

Documentation from other government and industry sources supports the position that smooth finished concrete floors are sufficiently nonsparking for the storage of fully packaged explosives.

Held, explosives storage facilities with smooth finished concrete floors that were constructed under contract for the use of the Department of Defense and that are presently being leased to licensees and permittees for the storage of commercial explosives are considered to be in compliance with the requirements for nonsparking floors, as set forth in 27 CFR 181.187(a)(4) and 181.187(b) and 27 CFR 181.90, for the storage of all types of fully packaged explosives, pyrotechnics and propellants, with the exception of black powder. Any other such magazines which have smooth finished concrete floors and which meet or exceed the Department of Defense construction specifications will also be considered to be in compliance with the requirements of Part 181 with respect to nonsparking floors. It is the responsibility of the licensee or permittee to provide verification that such facilities were manufactured under Department of Defense specifications or that the facilities meet or exceed such specification standards.

If it is determined by the Regional Director that the concrete floors of Type 1 or Type 4 explosives storage facilities do not meet the requirements as stated above, he will require such floors to be covered with a nonsparking material, such as epoxy paint or mastic.

Signed: June 26, 1975

[Editor's Note: 27 CFR Part 181 is now 27 CFR Part 555.]

3. 27 CFR 181.41: General (Also 181.11)

Certain companies that manufacture explosive materials for use in their own operations are required to obtain licenses as manufacturers of explosive materials.

ATF Ruling 75–31

The Bureau of Alcohol, Tobacco and Firearms has been asked whether certain companies, such as public utility companies that manufacture explosives for use in the area of field maintenance and construction activities, are "engaged in the business" as manufacturers of explosive materials and as such must obtain the required license for such manufacturers.

Under the provisions of 18 U.S.C. 842(a)(1), implemented by the regulations promulgated thereunder (27 CFR 181.41), it is unlawful to engage in the business of manufacturing explosives without a license. 18 U.S.C. 841(h) defines "manufacturer" as "any person engaged in the business of manufacturing explosive materials for purposes of sale or distribution or for his own use."

Although the term "engaged in the business" is not susceptible to a rigid definition, it is generally interpreted to imply an element of continuity or habitual practice as against a single act or occasional participation.

Accordingly, it is held that companies, such as public utility companies engaged in line and facility construction, which manufacture explosives on a regular or continual basis are considered to be engaged in the business of manufacturing explosive materials and must be appropriately licensed as required by 81 U.S.C. 842.

Signed: September 12, 1975

[Editor's Note: 27 CFR Part 181 is now 27 CFR Part 555.]

4. 27 CFR 555.109: Identification of Explosive Materials

Methods of marking containers of explosive materials are prescribed.

ATF Ruling 75–35

Editors note: ATF Ruling 75–35 was rendered obsolete pursuant to ATF 5F, 70 Federal Register 30626 (May 27, 2005), and effective July 26, 2005.

5. 27 CFR 55.11: Meaning of Terms—State of Residence

"State of residence" of business entities who use explosive materials; distribution of explosive materials by licensees to out-of-State business entities other than licensees and permittees; and distribution to nonresident employees of such entities are discussed.

Editor's Note: Provisions of ATF Ruling 76–4 were modified, in part, by the Safe Explosives Act.

Effective May 24, 2003, it is unlawful for any person to receive explosive materials unless such person holds an ATF license of permit. It is also unlawful for any licensee or permittee to knowingly distribute explosive materials to any person who does not hold a license or permit. The only relevance remaining in the term "State of residence" is for distribution of explosive materials to, and receipt by, limited permit holders. Pursuant to 18 U.S.C. 842(a)(3) and (a)(4), limited permit holders may, on not more than 3 separate occasions, lawfully receive explosive materials from a licensee or permittee whose premises are located within the state of residence of the limited permit holder. ATF Rule 76–4 continues to apply in determining whether a limited permit holder has acquired a "State of residence" for purposes of receipt of explosives under 18 U.S.C. 842(a)(4)(B).

6. 27 CFR 55.126: Explosives Transaction Record.

Under certain conditions, a single Form 5400.4 may be used to cover a series of deliveries.

Editor's Note: The provisions of ATF Rule 76–10 were rendered obsolete by ATF No. 1, 68 FR 13791, Mar. 20, 2003.

7. 27 CFR 181.187: Construction of Type 1 Storage Facilities (Also 181.188, 181.189)

Alternate construction standards for storage facilities for explosive materials are prescribed.

ATF Ruling 76–18

The Bureau of Alcohol, Tobacco and Firearms has reviewed the construction standards for storage facilities contained in Subpart J of 27 CFR Part 181 to determine if such construction criteria for bullet resistance meet current safety standards recognized by the explosives industry.

Under the provisions of 18 U.S.C. 842(j), it shall be unlawful for any person to store any explosive material in a manner not in conformity with regulations promulgated by the Secretary. In promulgating such regulations, the Secretary shall take into consideration the class, type, and quantity of explosive materials to be stored, as well as the standards of safety and security recognized in the explosives industry.

The regulations in 27 CFR 181.187, 181.188, and 181.189 prescribe types of storage facilities for explosive materials and provide, among other things, that such storage facilities shall be bullet resistant. 27 CFR 181.181(b) provides that alternate storage facilities may be authorized for the storage of explosive materials when it is shown that such alternate facilities are or will be constructed in a manner substantially equivalent to the standards of construction contained in the applicable regulations.

The term bullet-resistant means resistant to penetration of a bullet of 150 grain M2 ball ammunition having a nominal muzzle velocity of 2700 feet per second fired from a .30 caliber rifle from a distance of 100 feet perpendicular to the wall or door.

It has been determined that a wide range of construction criteria meet the bullet-resistant requirements of regulations for construction of storage facilities for explosive materials.

In order to promote standards of safety and security in the storage of explosive materials while allowing the industry a wide latitude in the selection of construction material, it is held that storage facilities (magazines) that are constructed according to the following minimum specifications are bullet resistant and meet the requirements of the regulations as set forth in 27 CFR Part 181. (All steel and wood dimensions indicated are actual thicknesses. To meet the concrete block and brick dimensions indicated, the manufacturer's represented thicknesses may be used.)

(a) Exterior of ⅝" steel, lined with an interior of any type nonsparking material.

(b) Exterior of ½" steel, lined with an interior of not less than ⅜" plywood.

(c) Exterior of ⅜" steel, lined with an interior of two inches of hardwood.

(d) Exterior of ⅜" steel, lined with an interior of three inches of softwood or 2¼" of plywood.

(e) Exterior of ¼" steel, lined with an interior of three inches of hardwood.

(f) Exterior of ¼" steel, lined with an interior of five inches of softwood or 5¼" of plywood.

(g) Exterior of ¼" steel, lined with an intermediate layer of two inches of hardwood and an interior lining of 1½" of plywood.

(h) Exterior of 3/16" steel, lined with an interior of four inches of hardwood.

(i) Exterior of 3/16" steel, lined with an interior of seven inches of softwood or 6¾" of plywood.

(j) Exterior of 3/16" steel, lined with an intermediate layer of three inches of hardwood and an interior lining of ¾" plywood.

(k) Exterior of ⅛" steel, lined with an interior of five inches of hardwood.

(l) Exterior of ⅛" steel, lined with an interior of nine inches of softwood.

(m) Exterior of ⅛" steel, lined with an intermediate layer of four inches of hardwood and an interior lining of ¾" plywood.

(n) Exterior of any type of fire-resistant material which is structurally sound, lined with an intermediate layer of four inches solid concrete block or four inches solid brick or four inches of solid concrete, and an interior lining of ½" plywood placed securely against the masonry lining.

(o) Standard eight inch concrete block with voids filled with well-tamped sand/cement mixture.

(p) Standard eight inch solid brick.

(q) Exterior of any type of fire-resistant material which is structurally sound, lined with an intermediate six inch space filled with well-tamped dry sand or well-tamped sand/cement mixture.

(r) Exterior of ⅛" steel, lined with a first intermediate layer of ¾" plywood, a second intermediate layer of 3⅝" well-tamped dry sand or sand/cement mixture and an interior lining of ¾" plywood.

(s) Exterior of any type of fire-resistant material, lined with a first intermediate layer of ¾" plywood, a second intermediate layer of 3⅝" well-tamped dry sand or sand/cement mixture, a third intermediate layer of ¾" plywood, and a fourth intermediate layer of two inches of hardwood or 14-gauge steel and an interior lining of ¾" plywood.

(t) Eight inch thick solid concrete.
Signed: June 23, 1976

[Editor's Note: 27 CFR Part 181 is now 27 CFR Part 555.]

8. 27 CFR 181.193: Quantity and Storage Restrictions (Also 181.188)

Alternate magazine construction standards for storage of electric blasting caps with other explosive materials are prescribed.

ATF Ruling 77–24

The Bureau of Alcohol, Tobacco and Firearms has been requested to authorize the storage of electric blasting caps in a separate compartment of a type two portable magazine.

Under the provisions of 18 U.S.C. 842(j), it shall be unlawful for any person to store any explosive material in a manner not in conformity with regulations promulgated by the Secretary. In promulgating such regulations, the Secretary shall take into consideration the class, type, and quantity of explosive materials to be stored, as well as the standards of safety and security recognized in the explosives industry.

Regulations in 27 CFR 181.193 restrict the storage of blasting caps with other explosive materials. Section 181.181(b) provides that alternate storage magazines may be authorized for the storage of explosive materials when it is shown that such alternate magazines are or will be constructed in a manner substantially equivalent to the standards of construction contained in the applicable regulations.

The Bureau recognizes that the transportation and the storage of explosive materials in the same vehicle along with electric blasting caps is often desired. The Institute of Makers of Explosives has established a recommended standard for such transport in their Safety Library Publication No. 22, dated November 5, 1971 and revised July 1976 [and further revised January 1985]. This standard prescribes the minimum construction criteria for (a) a container securely attached (1) above the cab of a vehicle (see Figure 1, Exhibit A), or (2) attached to the vehicle frame under the cargo compartment (see Figure 2, Exhibit A), or (b) a built-in compartment in the cargo space of the vehicle (see Exhibit B). In addition to motorized vehicles, consideration was also given for the use of similar criteria on portable wheeled trailers being used as magazines under section 181.188(a) of the regulations (see Exhibit E).

In order to insure standards of safety and security in the storage of explosive materials while allowing the industry a proper latitude in the construction of magazines, it is held that vehicles used for transporting and for storing explosive materials that are constructed in conformity with the standards listed below, and in compliance with all other safety and security provisions contained in Part 181, i.e., effectively immobilized when unattended, will meet the requirements of ATF regulations. Even though constructed on the same vehicle, each compartment will be considered as a separate magazine. The two magazines on the vehicle will, however, be considered as one magazine when applying the American Table of Distances.

Construction Standards for Storage of Electric Blasting Caps (Non Mass-Detonating)

a. The container or compartment must provide for total enclosure of the electric blasting caps.

b. The partition between the explosives storage compartment and the electric blasting cap compartment must be of laminate construction consisting of A/C grade or better exterior plywood, gypsum wall board [sheetrock] and low carbon steel plates. In order of arrangement, the laminate must conform to the following, with minimum thickness of each lamination as indicated: ½" plywood, ½" gypsum wall board [sheetrock] or ¼" asbestos board, ⅛" low carbon steel, and ¼" plywood, with the ¼" plywood facing the explosives storage compartment. See Exhibit C for details of laminate construction. The door to the electric blasting cap compartment must be of metal construction or solid wood covered with metal, the outside walls and top must be of the same construction as the rest of the vehicle or trailer. If high explosives, or bullet sensitive explosive materials are stored in the vehicle, then the storage compartment of the vehicle must be constructed so as to be bullet-resistant.

   c. As an alternative to the construction requirements shown in paragraph b, a container for use only as illustrated in Exhibit A may be used when constructed as follows:

     1. The top, lid or door, and the sides and bottom of each container must be of laminate construction consisting of A/C grade or better exterior plywood, solid hardwood, asbestos board and sheet metal. In order of arrangement, the laminate must consist of the following with the minimum thickness of each lamination as indicated: ¼" plywood, 1" solid hardwood, ½" plywood, ¼" asbestos board and 22-gauge sheet metal constructed inside to outside in that order. See Exhibit D for details of laminate construction.

     2. The hardwood must be fastened together with wood screws, the ½" plywood must be fastened to the hardwood with wood screws, the inner ¼" plywood must be fastened to the hardwood with adhesive and the 22-gauge sheet metal must be attached to the exterior of the container with screws.

   d. The laminate composite material must be securely bound together by waterproof adhesive or other equally effective means.

   e. The steel plates at the joints of laminations must be secured by continuous fillet welds.

   f. All interior surfaces of the container or compartment must be constructed so as to prevent contact of contents with any sparking metal.

   g. There must be direct access to the container or into a compartment from outside the vehicle.

   h. Each container or compartment must have a snug fitting continuous piano-type hinged lid or door equipped with a locking device/devices.

   i. Without permitting direct access to contents under normal conditions, the locking or hinging mechanisms must permit at least one edge of the lid or door to rise or move outward at least ½" when subjected to internal pressure.

   j. The exterior of the container or compartment must be weather-resistant.

Signed: June 3, 1977

[Editor's Note: 27 CFR Part 181 is now 27 CFR Part 555.]

9. 18 U.S.C. 842(j): Storage of Explosives

27 CFR 55.208(b)(1), 55.210(b)(1), and 55.211(b)(1): Indoor Storage of Explosives in a Residence or Dwelling

ATF will approve variances to store explosives in a residence or dwelling only upon certain conditions including, but not limited to, receipt of a certification of compliance with State and local law, and documentation that local fire safety officials have received a copy of the certification.

ATF Ruling 2002–3

The Bureau of Alcohol, Tobacco and Firearms (ATF) has received questions concerning indoor storage of explosives in a residence or dwelling and whether such storage must comply with State or local law.

Section 842(j) of 18 U.S.C. states: "It shall be unlawful for any person to store any explosive material in a manner not in conformity with regulations promulgated by the Secretary."

The regulations in 27 CFR 55.208(b)(1), 55.210(b)(1), and 55.211(b)(1) specify that no indoor magazine is to be located in a residence or dwelling. Section 55.22 specifies that the Director may allow alternate methods or procedures in lieu of a method or procedure specifically prescribed in the regulations. Specifically, section 55.22(a)(3) provides that such "variances" are permissible only in certain circumstances, including where "[t]he alternate method or procedure will not be contrary to any provision of law and will not…hinder the effective administration of this part."

ATF has been advised that certain variances previously approved for storage of explosives in residences or dwellings are in violation of State or local zoning law. ATF believes it is important to ensure that approval of variances is in compliance with all State and local provisions.

To obtain a variance for indoor storage of explosives in a residence or dwelling, ATF has determined that a person must submit to ATF a certification signed under penalty of perjury along with the request for the variance. The certification must:

   1. State that the proposed alternative storage method will comply with all applicable State and local law;

   2. Provide the name, title, address, and phone number of the authority having jurisdiction for fire safety of the locality in which the explosive materials are being stored; and,

   3. Demonstrate that the person has mailed or delivered the certification to the authority identified in (2).

When required by the Director, such persons must furnish other documentation as may be necessary to determine whether a variance should be approved.

Held, ATF will approve variances to store explosives in a residence or dwelling only upon certain conditions including, but not limited to, receipt of a certification of compliance with State and local law, and documentation that local fire safety officials have received a copy of the certification.

Date signed: August 23, 2002

10. 18 U.S.C. 842(j): Storage of Explosives

27 CFR 55.208(b)(1), 55.210(b)(1), and 55.211(b)(1): Indoor Storage of Explosives in Business Premises Directly Adjacent to a Residence or Dwelling

ATF requires approval of variances for indoor storage of explosives in business premises directly adjacent to a residence or dwelling.

ATF Ruling 2002–4

The Bureau of Alcohol, Tobacco and Firearms (ATF) has received questions concerning indoor storage of explosives in business premises directly adjacent to a residence or dwelling.

Section 842(j) of 18 U.S.C. states: "It shall be unlawful for any person to store any explosive material in a manner not in conformity with regulations promulgated by the Secretary."

The regulations in 27 CFR 55.208(b)(1), 55.210(b)(1), and 55.211(b)(1) specify that no indoor magazine is to be located in a residence or dwelling. Section 55.22 specifies that the Director may allow alternate methods or procedures in lieu of a method or procedure specifically prescribed in the regulations. Specifically, section 55.22(a)(3) provides that such "variances" are permissible only in certain circumstances, including where "[t]he alternate method or procedure will not be contrary to any provision of law and will not…hinder the effective administration of this part."

ATF has been asked whether businesses that are directly adjacent to living quarters may lawfully store explosive materials in the business premises. The issue presented is whether the premises amount to a "residence or dwelling" within the meaning of the regulations cited above.

Even where the business premises are segregable from the living quarters by the existence of a door or a common wall, the business premises retain their character as a residence or dwelling. Accordingly, indoor storage of explosives in such premises is generally prohibited and can be allowed only pursuant to an approved variance.

Held, ATF requires approval of variances for indoor storage of explosives in business premises directly adjacent to a residence or dwelling. ATF may approve such variances upon receipt of all appropriate certification and other documentation as may be requested.

Date signed: August 23, 2002

11. 18 U.S.C. 842(f): Unlawful Acts

27 CFR 555.105(b)(6)(iii): Distribution of Explosives to Limited Permittees

Distributors distributing explosive materials to holders of limited permits via common or contract carrier may verify receipt of the explosive materials by telephone, facsimile, e-mail, or other means within three business days of shipment in lieu of requiring the common or contract carrier to verify the identity of the person accepting delivery of the explosives. The distributor shall make a notation on ATF Form 5400.4 indicating whether the shipment was received and the date and time of the contact with the distributee.

ATF Ruling 2003–5

The Bureau of Alcohol, Tobacco, Firearms and Explosives (ATF) has received questions from the explosives industry regarding the requirement under 27 CFR 555.105(b)(6)(iii) that, effective May 24, 2003, a common or contract carrier hired by a Federal explosives licensee or permittee verify the identity of the person accepting delivery on behalf of the distributee, note the type and number of the identification document, and provide this information to the distributor. The distributor is required to record this information on ATF Form 5400.4, Limited Permittee Transaction Report (LPTR).

Industry members have informed ATF that this requirement places an undue burden on common and contract carriers. Drivers are concerned that verifying the identity of persons accepting delivery of explosive materials by examining an identification document and providing the identification information to the distributor will be overly time consuming. Drivers are also concerned that they could be held personally liable for delivering explosives to persons not authorized to receive them.

ATF imposed the verification requirements of section 555.105(b)(6)(iii) to ensure that when explosive materials are sold by a distributor to a holder of a limited permit and transported by a common or contract carrier hired by the distributor, the explosive materials are delivered only to a person authorized to receive them. ATF continues to believe it is important that sellers of explosive materials verify that such materials are delivered to persons authorized to receive them. However, it was not ATF's intention to impose an undue burden on common or contract carriers.

Section 555.22, Title 27, CFR, provides that the Director may approve an alternate method or procedure in lieu of a method or procedure specifically prescribed in the regulations when he finds that:

(1) Good cause is shown for the use of the alternate method or procedure;

(2) The alternate method or procedure is within the purpose of, and consistent with the effect intended by, the specifically prescribed method or procedure and that the alternate method or procedure is substantially equivalent to that specifically prescribed method or procedure; and

(3) The alternate method or procedure will not be contrary to any provision of law and will not result in an increase in cost to the Government or hinder the effective administration of Part 555.

ATF finds that there is good cause to authorize a variance to the provisions of section 555.105(b)(6)(iii) due to the undue burden placed on common or contract carriers by the verification requirement. Accordingly, ATF authorizes the following alternate method or procedure to the identification verification requirements of section 555.105(b)(6)(iii):

The distributor shall, no later than three business days after shipment of the explosive materials, contact the distributee by telephone, facsimile, e-mail or any other means to ensure that the shipment has been received. The distributor shall make a notation on ATF Form 5400.4 indicating whether the shipment was received and the date and time of the contact with the distributee.

ATF finds that the above alternate method is consistent with the verification provisions of section 555.105(b)(6)(iii), because it will ensure that delivery has taken place and document the information in the distributor's records. The alternate method is not contrary to any provision of law, will not increase the costs to ATF, and will not hinder the effective administration of the regulations in 27 CFR Part 555.

Held, pursuant to 27 CFR 555.22, ATF authorizes a variance from the requirements of 27 CFR 555.105(b)(6)(iii) for Federal explosives licensees and permittees making distributions of explosive materials to holders of limited permits via common or contract carrier. As an alternate method or procedure, distributors distributing explosive materials to holders of limited permits via common or contract carrier may verify receipt of the explosive materials by telephone, facsimile, e-mail or other means within three business days of shipment in lieu of requiring the common or contract carrier to verify the identity of the person accepting delivery of the explosives. The distributor shall make a notation on ATF Form 5400.4 indicating whether the shipment was received and the date and time of the contact with the distributee.

Date signed: May 23, 2003.

12. 18 U.S.C. 842(j): Storage of Explosives 27 CFR 555.210(b)(4): Locking Requirements for Indoor Type 4 Storage Magazines

27 CFR 555.22: Alternate Methods or Procedures; Emergency Variations from Requirements

Under certain conditions, flush-mounted bolt-type locks will be considered adequate for locking type 4 indoor magazines

ATF Ruling 2004–3

The Bureau of Alcohol, Tobacco, Firearms and Explosives (ATF) has received inquiries from explosives industry members as to the suitability of certain types of locks for type 4 indoor storage magazines utilized for the storage of low explosives.

Section 842(j) of 18 U.S.C. states, "It shall be unlawful for any person to store any explosive material in a manner not in conformity with regulations promulgated by the Attorney General."

The regulations at 27 CFR 555.210(b) state, in part, "Indoor magazines are to be fire-resistant and theft-resistant." To satisfy the theft-resistance requirement, this section requires that each door be equipped with two mortise locks, two padlocks fastened in separate hasps and staples; a combination of a mortise lock and padlock; a mortise lock that requires two keys to open; or a three-point lock. Padlocks must have at least five tumblers and a case-hardened shackle of at least ⅜" diameter. In addition, padlocks must be protected with not less than ¼" steel hoods constructed so as to prevent sawing or lever action on the locks, hasps, and staples.

This section further provides, "Indoor magazines located in secure rooms that are locked as provided in this subparagraph may have each door locked with one steel padlock (which need not be protected by a steel hood) having at least five tumblers and a case-hardened shackle of at least ⅜" diameter, if the door hinges and lock hasp are securely fastened to the magazine." This section makes it clear that, although the magazine-locking standards are reduced for magazines secured in a locked room, a substantial locking mechanism is still required.

The regulations at 27 CFR 555.22 allow for the approval and use of an alternate method or procedure (variance), provided that (1) there is good cause for the proposed variance; (2) the proposed variance is consistent with and substantially equivalent to the prescribed method or procedure; and (3) the proposed variance is not contrary to any provision of law and will not hinder the effective administration of the regulations or result in an increase in cost to the Government.

ATF has been asked whether a cam-type lock known as a "flush-mount lever lock," or a similarly mounted lock with a bolt-type locking mechanism meets ATF theft-resistance requirements for type 4 indoor storage of low explosives.

ATF has determined that, although the lever-type lock is mounted on a magazine lid in such a manner as to preclude prying or cutting of the lock, the lever-locking mechanism does not provide adequate protection against pulling or prying the lid off the magazine. The lever mechanism rests under a small piece of metal on the edge of the magazine wall (typically 18-gauge sheet metal), to secure the lid. This type of lock fails to provide a level of theft-resistance for indoor storage of low explosive materials that would be substantially equivalent to the methods prescribed in the regulations.

ATF has also examined the flush-mounted bolt-style locks, which secure the magazine by means of a bolt-type mechanism. The cylinder portion of the lock mounts in the lid of the magazine in such a manner that, when the key is turned, the bolt slides toward the outer wall of the magazine. This bolt engages in a slotted locking block attached securely to the inside of the magazine wall. Because this locking mechanism relies upon interlocking solid metal parts, operating in a fashion similar to a deadbolt lock, it provides a level of theft resistance that is substantially equivalent to that required by the regulations.

Held, cam-type locks known as "flush-mount lever locks," or a similarly mounted lock with a bolt-type locking mechanism does not provide a level of theft-resistance for indoor storage of materials that is substantially equivalent to the methods prescribed in the regulations.

Held further, flush mount bolt-style locks utilizing interlocking solid metal parts, each affixed securely to the magazine in such a way that they cannot be readily removed from the exterior of the magazine and each locking mechanism having at least five tumblers, will be considered to meet the theft-resistance requirements of Part 555, section 210(b) for indoor type 4 explosives storage magazines.

Date approved: June 5, 2004

13. 27 CFR 555.11: Meaning of Terms

ATF provides guidance on three different private roads and whether they are "highways" as defined in 27 CFR 555.11.

ATF Ruling 2005–2

The Bureau of Alcohol, Tobacco, Firearms and Explosives (ATF) has received inquiries from members of the explosives industry as to the meaning of the term "highway" under 27 CFR 555.11.

The Federal explosives laws, 18 U.S.C. Chapter 40, require all persons to store explosive materials in a manner in conformity with regulations issued by the Attorney General. 18 U.S.C. 842(j).

The Attorney General has delegated his authority to administer and enforce the Federal explosives laws to the Director, ATF. 28 CFR 0.130. Regulations in 27 CFR Part 555 implement the provisions of the Federal explosives laws. Storage regulations in 27 CFR Part 555, Subpart K, provide that outdoor magazines in which high explosives are stored must be located no closer to inhabited buildings, passenger railways, public highways, or other magazines in which high explosives are stored than the minimum distances specified in the table of distances for storage of explosive materials in section 555.218 of the regulations. 27 CFR 555.206.

Section 555.11 of the regulations defines the term "highway" as "[a]ny public street, public alley, or public road, including a privately financed, constructed, or maintained road that is regularly and openly traveled by the general public."

In Scenario A, a private road owned by a corporation is used by the general public as an access road to a parking lot owned by the corporation. The road is near an explosives magazine. The road does not have a gate, sign, or other means of restricting access to the road. The road is also used by the general public on a daily basis to gain access to other public streets.

In Scenario B, a company that manufactures display fireworks, a logging company, and an individual who owns buildings utilized to store his collection of automobiles all occupy property to which the only access is a privately owned road. A separate party that leases to these three entities owns the property. The road is located on private property, and a locked gate at the entrance to the road prevents access by the general public. The display fireworks company, the logging company, and the individual storing automobiles all have keys to unlock the gate and travel on the road when needed. The gate is locked at all times, and there is no evidence that the road is open to anyone other than the two businesses and one individual who require access to their property.

In Scenario C, an explosives company maintains explosives magazines in a quarry area that has a roadway traversing through the quarry. The quarry owns the property, and the road is maintained by the quarry. The road has a gate and there are signs advising no trespassing. However, when ATF officials visited the location on several occasions, the gate was left open and members of the public regularly utilized the roadway as a shortcut between two major highways. There were no indications the owner of the property took any steps to prevent members of the public from utilizing the roadway.

Applying the regulatory definition of "highway" to the three scenarios, the road in Scenario A is clearly a highway that is subject to the tables of distance in Part 555. Although it is privately owned, it is regularly and openly traveled by members of the general public without restriction.

The roadway in Scenario B is not a "highway" as defined. Access is restricted at all times and there is no evidence the general public regularly travels on the roadway. Access to the road is provided to only a limited number of persons who have a legal right to travel the road. Accordingly, this road is not regularly and openly traveled by members of the general public.

ATF concludes that the roadway described in Scenario C is a "highway" as defined in 27 CFR 555.11. Although access to the roadway is restricted by a gate and "No trespassing" signs are posted, the gate is not closed at all times. Furthermore, ATF observation indicates that the roadway is regularly and openly traveled by members of the general public. Based on these facts, the roadway is a highway which is subject to the tables of distance in 27 CFR Part 555.

Held, a private road with no gate, signs, or other means of restricting access that is used by the general public as an access road to a parking lot and as access to other public streets is a "highway" as defined in 27 CFR 555.11.

Held further, a private road with a locked gate at the entrance that is locked at all times and used by a limited number of persons leasing or owning property accessed by the road is not a "highway" as defined in 27 CFR 555.11.

Held further, a private roadway traversing a quarry with a gate restricting access and a "no trespassing" sign is a "highway," as defined in 27 CFR 555.11, because the gate is not locked at all times and the general public regularly utilizes the roadway as a shortcut between two public highways.

Date approved: September 8, 2005

14. 27 CFR 555.11: Meaning of Terms

ATF provides guidance on two situations involving structures and whether they are "inhabited buildings" as defined in 27 CFR Part 555.

ATF Ruling 2005–3

The Bureau of Alcohol, Tobacco, Firearms and Explosives (ATF) has received inquiries from explosives industry members as to the meaning of the term "inhabited building" under 27 CFR 555.11.

The Federal explosives laws, 18 U.S.C. Chapter 40, require all persons to store explosive materials in a manner in conformity with regulations issued by the Attorney General. 18 U.S.C. 842(j). The Attorney General has delegated his authority to administer and enforce the Federal explosives laws to the Director, ATF. 28 CFR 0.130. Regulations in 27 CFR Part 555 implement the provisions of the Federal explosives laws. Storage regulations in 27 CFR Part 555, Subpart K, provide that outdoor magazines in which high explosives are stored must be located no closer to inhabited buildings, passenger railways, public highways, or other magazines in which high explosives are stored than the minimum distances specified in the table of distances for storage of explosive materials in section 555.218 of the regulations. 27 CFR 555.206.

The regulation at 27 CFR 555.11 defines the term "inhabited building" as "[a]ny building regularly occupied in whole or in part as a habitation for human beings, or any church, schoolhouse, railroad station, store, or other structure where people are accustomed to assemble, except any building occupied in connection with the manufacture, transportation, storage, or use of explosive materials."

In Scenario A, an explosives licensee leases explosives magazines to an individual who uses the magazines for storage of goods other than explosive materials. The magazines are located adjacent to magazines used by the explosives licensee for the storage of explosive materials. The magazines used by the lessee are not separated by the minimum distances required for the separation of magazines from "inhabited buildings" as required by the regulations in 27 CFR Part 555.

The magazines are visited regularly by the individual who stores property in the magazines, but no additional persons accompany the individual when he is present at the magazine. However, the individual hires contractors to repair equipment stored in the facilities from time to time, and 1–3 employees of the contractor may occasionally be present for short periods of time at the storage site. However, such visits occur no more than 3–5 times per year. In addition, an employee from the water company visits the storage site once a month to read the water meter, an employee from the power company reads the power meter once a month, and other vendors may be present at the site for short periods of time for other legitimate purposes.

In Scenario B, a licensed manufacturer of explosives X leases a unit in an industrial park that shares a common wall with a unit leased by licensed manufacturer Y. Both licensees store explosives in magazines located inside and outside the units. The magazines of Manufacturer X and the building used by Manufacturer Y are not separated by the minimum distance prescribed in 27 CFR Part 555 for the separation of magazines and inhabited buildings. Likewise, the magazines of Manufacturer Y and the building used by Manufacturer X are not separated by the minimum distance prescribed in 27 CFR Part 555 for the separation of magazines and inhabited buildings. In both structures, employees and contractors are regularly present during work hours for purposes of carrying on the manufacturing and distributing businesses of the two licensees. This includes personnel who work in the manufacturing plant, those who work on the loading dock to load and ship explosives products to customers, and those who work in the office taking orders, sending out invoices, and handling other clerical work for the businesses.

Applying the law and regulations to the facts of Scenario A, ATF concludes that the leased structures used by the individual to store items other than explosives are not being used as a habitation for human beings and are not buildings occupied in connection with the manufacture, transportation, storage, or use of explosive materials. Accordingly, the sole issue remaining is whether the structure is one where people are accustomed to assemble.

Noteworthy, the regulation uses the term "people," which is the plural version of "person." Thus, ATF believes that more than one person must "assemble" at the structure for it to be an "inhabited building." In addition, the word "assemble" is defined, in part, as "To bring or gather together in a group or whole." The word "assembly" is defined, in part, as "A group of persons gathered for a common purpose." The American Heritage Dictionary, Second College Edition, Houghton Mifflin Co., 1982. It is clear that the presence of one person at a structure or location cannot be an assembly of any sort. Accordingly, in situations where one person is present at a particular structure, whether on a regular or infrequent basis, such a structure is not an "inhabited building" as defined in 27 CFR 555.11.

Likewise, occasional visits to the storage facility by mail delivery persons or employees of public utility companies for brief periods of time would not be an "assembly" that would make the facility an inhabited building. However, if 2 or more repair persons are present at the facility to make repairs to equipment stored there, such persons would be there for a common purpose, and would have "assembled" at the structure. However, the structure would be an "inhabited building" only if it is a structure where people are accustomed to assemble. The word "accustom" is defined as "To familiarize, as by constant practice, use, or habit: accustomed himself to working long hours." The word *"accustomed"* is defined as "Usual, characteristic, or normal: worked with her accustomed thoroughness." The American Heritage Dictionary, Second College Edition, Houghton Mifflin Co., 1982. These definitions indicate that a structure will be one where people are *accustomed* to assemble only if there is some degree of continuity, regularity, or frequency to such assembly.

Infrequent, occasional visits to the storage site by 2 or more repair persons would not make the storage facility an "inhabited building," because such intermittent visits would not be "customary." Only where 2 or more persons are present at the site for a common purpose and on a regular basis would the building fit within the definition of "inhabited building."

To address Scenario B, it is apparent that the building leased by Manufacturer X is exempt from the definition of "inhabited building" as to the magazines of Manufacturer X, and the building leased by Manufacturer Y is likewise exempt as to the magazines of Manufacturer Y. This is because both buildings are occupied in connection with the manufacture, transportation, and storage of explosive materials. Clearly, the employees of both licensees are aware that explosive materials are present on the premises and they assume the risk of any such operation. A more difficult question is presented by the buildings of Manufacturer X and the magazines of Manufacturer Y and vice versa. ATF cannot assume that all employees are cognizant of the activities of their neighbors in the industrial park. Thus, it cannot be assumed that the employees are knowingly assuming the risk of explosive materials stored in magazines owned by the other licensee.

Given the plain language of the regulation, however, ATF does not believe it is appropriate to deny the coverage of the regulatory exemption to adjoining licensees on the basis of an assumption of the risk analysis. The current regulatory definition excludes from the definition of "inhabited building" any building occupied in connection with the manufacture, transportation, storage, or use of explosive materials, regardless of the knowledge of the building's occupants. Accordingly, ATF concludes that the industrial units occupied by Manufacturer X and Manufacturer Y are both exempted from the definition of "inhabited building" as to the magazines of each other as well as to their own magazines.

Held, a structure used to store items other than explosive materials that is visited on a regular basis by one individual is not an "inhabited building" as defined in 27 CFR 555.11, because it is not a structure where people are accustomed to assemble. Where 2 or more persons are present at the structure to repair equipment stored therein and such visits to the site are occasional and infrequent, the structure is not an "inhabited building" because the visits are not "customary." However, where 2 or more persons make regular visits to the structure for a common purpose, the structure is an "inhabited building," and explosives magazines may not be stored closer to the structure than the minimum distances specified in the regulations in 27 CFR Part 555.

Held further, buildings occupied by licensed explosives manufacturers in connection with the manufacture, transportation, storage, or use of explosive materials are not included within the definition of "inhabited building" as to magazines located on their own premises. In addition, buildings occupied by licensed explosives manufacturers in connection with the manufacture, transportation, storage, or use of explosives are not included within the definition of "inhabited building" as to magazines located on property owned by another licensee.

Date approved: November 25, 2005

15. 18 U.S.C. 842(f): Records Required for Explosives Licensees and Permittees

27 CFR PART 555, Subpart G: Records and Reports

27 CFR 555.22: Alternate Methods or Procedures; Emergency Variations from Requirements

Under specified conditions, approval is granted to utilize computerized records as required records under 27 CFR 555, Subpart G.

ATF Ruling 2007–1

The Bureau of Alcohol, Tobacco, Firearms and Explosives (ATF) has received inquiries from members of the explosives industry about maintaining their required Federal explosives records in computerized form rather than paper form.

Section 842(f), Title 18, United States Code, makes it unlawful for any licensee or permittee to willfully manufacture, import, purchase, distribute, or receive explosive materials without making such records as the Attorney General may by regulation require, including, but not limited to, a statement of intended use, the name, date, place of birth, social security number or taxpayer identification number, and place of residence of any natural person to whom explosive materials are distributed.

Regulations implementing section 842(f) are in 27 CFR Part 555, Subpart G. The regulations in this subpart specify the records required to be created and maintained by licensed importers (section 555.122), licensed manufacturers (section 555.123), licensed dealers (section 555.124), and permittees (section 555.125). The regulation in section 555.121 provides that licensees and permittees must keep records pertaining to explosive materials in permanent form (i.e., commercial invoices, record books) and in the manner required in Subpart G. In addition, sections 555.122–555.125 specifically allow licensees and permittees to use an alternate record to record the distribution of explosive materials when it is shown that the alternate records would accurately and readily disclose the information required by the regulations. These regulations require licensees and permittees who propose to use alternate records to submit a letter application to ATF describing the proposed alternate records and the need for them. Alternate records are not to be employed until approval from ATF is received.

Regulations at 27 CFR 555.22 allow for the approval and use of an alternate method or procedure in lieu of a method or procedure specifically prescribed in Part 555. ATF may approve an alternate method or procedure when it is found that—

(1) Good cause is shown for the use of the alternate method or procedure;

(2) The alternate method or procedure is within the purpose of, and consistent with the effect intended by, the specifically prescribed method or procedure and that the alternate method or procedure is substantially equivalent to that specifically prescribed method or procedure; and

(3) The alternate method or procedure will not be contrary to any provision of law and will not result in an increase in cost to the Government or hinder the effective administration of 27 CFR Part 555.

With advances in technology and the dramatic decrease in the cost of computers, many businesses rely upon computers to maintain records of their inventory, sales, customer lists, and other business information. Even the smallest home-based business utilizes computers to record and maintain business information. Creating and maintaining records in a computer database, rather than paper form, makes it easier to ensure accuracy of records and makes it less likely that records will be lost or misplaced. In addition, maintaining records via computer generally saves time and money in bookkeeping and auditing expenses. This utilization of computers has allowed companies to automate inventories, utilizing technology such as bar codes or RFID (radio frequency identification) chips. This facilitates better accountability of product overall, reducing the potential of everyday accounting errors. Over the years ATF has seen a significant increase in the number of requests from explosives licensees and permittees for authorization to utilize computerized records rather than paper records of acquisition and distribution and other required records, such as magazine transaction records. ATF routinely approves requests to utilize computerized records, with certain conditions, finding that the use of such records is substantially equivalent to methods of record keeping set forth in the regulations in 27 CFR Part 555, Subpart G.

Several explosives industry members have asked whether computerized records may be maintained without obtaining written approval from ATF if they contain all the required information specified in the regulations and are maintained in a permanent form. Additionally, industry members have questioned whether computer records in combination with paper records may be maintained if they are permanent and contain all the information required by the regulations.

ATF has determined that records of acquisition and disposition, magazine summary records, and the other records required by 27 CFR Part 555, Subpart G, satisfy the standard of permanency and are substantially equivalent to paper records if they meet the following criteria:

1. All data entered into the computer system must be recorded into the database and cannot be capable of being edited or modified at a later date. The software system must retain any correction of errors as an entirely new entry, without deleting or modifying the original entry. The system may allow for entries in a notes column to explain any correction.

2. The system must have a reliable daily memory backup capability to protect the data from accidental deletion or other system failure.

It is also acceptable for licensees/permittees to maintain required records using a combination of a computer program, commercial invoices, and other documents, provided that all of the required information is maintained in the records in permanent form. Any use of a computer for any portion of the required records must comply with the standards outlined above. However, each particular transaction must be self-contained with all the required information in the same recordkeeping medium. As one example, dispositions of explosives by a dealer cannot be separated by keeping the dates of disposition and the manufacturer's name or brand name in the computer, and all the other required information for that disposition on separate written documents.

ATF finds that good cause exists for authorizing the use of a computer to create and maintain the records required by 27 CFR Part 555, Subpart G, as the use of computers is accepted throughout the business community as a reliable, cost-efficient means of maintaining business records. ATF also finds that the use of a computer to maintain required records, contingent upon the requirements outlined above, is consistent with the effect intended by the requirements of Subpart G, as it will result in a permanent, reliable record that will accurately indicate acquisitions and dispositions of explosive materials. Finally, ATF finds that the use of computer records properly containing all the required information should not hinder the effective administration of the Federal explosives laws or regulations – use of such records generally makes it easier for ATF to conduct inventories of product on hand and to audit required records. Accordingly, ATF concludes that the requirements for approval of an alternate method or procedure in accordance with 27 CFR Part 555, sections 555.22 and 555.122–555.125, are met.

Held, persons holding licenses and permits issued under 18 U.S.C., Chapter 40, may use computers to create and maintain all or any portion of the records required by 18 U.S.C. 842(f) and 27 CFR Part 555, Subpart G, if the following conditions are satisfied:

1. All data entered into the computer system must be recorded into the database and cannot be capable of being edited or modified at a later date. The software system must retain any correction of errors as an entirely new entry, without deleting or modifying the original entry. The system may allow for entries in a notes column to explain any correction.

2. The system must have a reliable daily memory backup capability to protect the data from accidental deletion or other system failure.

Held further, licensees and permittees who wish to use computers to create and maintain all or a portion of their required records in accordance with the requirements set forth in this ruling are not required to obtain advance approval in accordance with 27 CFR 555.22 or 555.121–125.

84

Held further, licensees and permittees utilizing a combination of a computer program, commercial invoices, and other paper documents as required records must ensure that the required information for a particular transaction is fully contained in the same recordkeeping medium.

Held further, this ATF approved alternate method or procedure for computerized records shall not be withdrawn unless the holder of said variance is so advised by ATF in writing or no longer holds a Federal explosives license or permit.

Date approved: January 18, 2007

16. 18 U.S.C. 842(j): Storage of Explosives

27 CFR 555.210: Construction of Type 4 Magazines

27 CFR 555.215: Housekeeping

27 CFR 555.201: Notification of Local Fire Officials

27 CFR 555.63: Explosives Magazine Changes

27 CFR 555.22: Alternative Methods or Procedures; Emergency Variations from Requirements

Under specified conditions, display fireworks may be temporarily stored in locked and attended motor vehicles at the explosives magazine site(s) and at fireworks display site(s) without meeting the locking requirements of 27 CFR 555.210 provided certain additional security measures are in place. Additionally, allowance per 27 CFR 555.215 is made for the fuel tanks containing volatile materials that may be on the temporary storage vehicles. Finally, slight variation is provided for notification requirements to ATF and local fire officials.

ATF Ruling 2007–2

The Bureau of Alcohol, Tobacco, Firearms and Explosives (ATF) has received inquiries from members of the explosives industry concerning the necessity of applying for a variance each year to cover the temporary storage of display fireworks at explosives magazine storage site(s) and fireworks display site(s). The continual reapplication and issuance of these variances is a burden on both industry and the Government with little or no benefit to safety or security.

The Federal explosives laws, 18 U.S.C. Chapter 40, require all persons to store explosive materials in a manner in conformity with regulations issued by the Attorney General. 18 U.S.C. 842(j). The Attorney General has delegated the authority to administer and enforce the Federal explosives laws to the Director, ATF. 28 CFR 0.130. Regulations in 27 CFR Part 555, implement the provisions of the Federal explosives laws.

The regulation at 27 CFR 555.210(a) states, in part, "[o]utdoor magazines are to be fire-resistant, weather-resistant, and theft-resistant." Partly to satisfy the theft-resistant requirement, this section requires that each door be equipped with two mortise locks; two padlocks fastened in separate hasps and staples; a combination of a mortise lock and a padlock; a mortise lock that requires two keys to open; or a three-point lock. Padlocks must have at least five tumblers and a case-hardened shackle of at least ⅜" diameter.

Padlocks must be protected with not less than ¼" steel hoods constructed so as to prevent sawing or lever action on the locks, hasps, and staples.

The regulation at 27 CFR 555.215 states, in part, "[v]olatile materials are to be kept a distance of not less than 50 feet from outdoor magazines."

The regulation at 27 CFR 555.201 requires, in part, that any person storing explosive materials notify local fire authorities orally before the end of the day on which storage of the explosive materials began and in writing within 48 hours from the time such storage began.

The regulation at 27 CFR 555.63 requires that any licensee or permittee who acquires (adds) a storage magazine must notify ATF at least five business days in advance of using any added explosives storage magazine.

Regulations at 27 CFR 555.22 allow for the approval and use of an alternate method or procedure in lieu of a method or procedure specifically prescribed in Part 555. ATF may approve an alternate method or procedure when it is found that—

(1) Good cause is shown for the use of the alternate method or procedure;

(2) The alternate method or procedure is within the purpose of, and consistent with the effect intended by, the specifically prescribed method or procedure and that the alternate method or procedure is substantially equivalent to that specifically prescribed method or procedure; and

(3) The alternate method or procedure will not be contrary to any provision of law and will not result in an increase in cost to the Government or hinder the effective administration of 27 CFR Part 555.

ATF has approved a significant number of variances for temporary storage for a specified amount of time before a display fireworks event, as well as during and after the event until the remaining explosive materials can be placed back into the appropriate storage magazine.

Preparation of display fireworks shows and the transportation of explosive materials to numerous show sites often take place over a period of several days. Preparing and temporarily storing the fireworks for these shows ordinarily take place on delivery trucks and trailers in one storage location where the proprietor already maintains storage of explosives materials with a high degree of security and safety by complying with the provisions of 27 CFR Part 555.

Many display fireworks shows also take several days to prepare at the show site. During preparation and after the show is completed, explosive materials frequently must be temporarily stored. This is often either extra product that was brought to the show or misfires that have been maintained and must be returned to permanent storage.

Allowing flexibility through alternate methods or procedures for specific regulations increases both safety and security at these show sites. These procedures are needed to increase public safety, as well as facilitate smooth operations for the display fireworks industry. The highest risk of incidents involving the accidental ignition of display fireworks is during handling, with the next highest risk being transportation. Providing no flexibility to allow storage in the delivery vehicles would require the industry to dangerously load and unload from storage magazines to vehicles and back into a storage magazine. Additionally, ATF believes that providing this guidance allows for preplanning by the proprietor and consistency of regulatory application nationwide.

One of the major dangers around explosives is fire. Therefore, the regulations require that volatile materials be maintained a distance of not less than 50 feet from outdoor explosives storage magazines. ATF believes that requiring attended storage for display fireworks temporarily stored in vehicles will ensure public safety, in lieu of the 50 foot separation requirement. The attendee should be able to alert the proper authorities if needed to ensure that a fire does not compromise this storage, or may relocate these temporary storage magazines to a safe location away from an identified fire.

Held, ATF will approve alternate methods or procedures for the temporary storage of display fireworks in locked and attended vehicles at explosives magazine site(s), as well as at the fireworks display site(s), under the following conditions:

1. The doors to each storage compartment containing explosive materials must be locked with at least one steel padlock having at least five tumblers and a casehardened shackle of at least ⅜" diameter. The padlock does not need to be protected by a steel hood. However, each temporary storage magazine must be attended at all times for security purposes. The vehicle is considered "attended" when an authorized individual is within 100 feet of all temporary storage and has an unobstructed view of the vehicle(s) containing the explosive materials. The individual must remain awake and observant of activities around the vehicle(s).

2. The person who temporarily stores the explosive materials must notify in writing the authority having jurisdiction for fire safety in the locality in which the explosive materials are stored no less than 3 Federal office business days prior to utilizing the additional temporary storage magazine(s).

3. The person who temporarily stores the explosive materials must notify ATF in writing of the location of this storage no less than 3 Federal office business days prior to utilizing the additional temporary storage magazine(s).

All other provisions of 27 CFR Part 555 must be complied with as prescribed.

Held further, this ATF-approved alternate method or procedure for the temporary storage of display fireworks in locked and attended vehicles shall not expire unless the holder of said variance is so advised by ATF or no longer holds a Federal explosives license or permit.

Date approved: January 18, 2007

17. 18 U.S.C. 842(j): Storage of Explosives

27 CFR 555.211(a): Immobilization of Outdoor Type 5 Mobile Storage Magazines

27 CFR 555.215: Housekeeping

27 CFR 555.22: Alternate Methods or Procedures; Emergency Variations from Requirements

Under specified conditions, blasting agents may be stored in mobile type 5 magazines (bulk delivery trucks) without meeting the prescribed immobilization requirements of 27 CFR 555.211.

ATF Ruling 2007–3

The Bureau of Alcohol, Tobacco, Firearms and Explosives (ATF) has received inquiries from members of the explosives industry concerning the preloading and temporary storage of blasting agents on bulk delivery vehicles.

The Federal explosives laws, 18 U.S.C. Chapter 40, require all persons to store explosive materials in a manner in conformity with regulations issued by the Attorney General. 18 U.S.C. 842(j). The Attorney General has delegated his authority to administer and enforce the Federal explosives laws to the Director, ATF. 28 CFR 0.130. Regulations in 27 CFR Part 555 implement the provisions of the Federal explosives laws.

The regulation at 27 CFR 555.211(a)(1) states, in part, "[o]utdoor magazines are to be weather-resistant and theft-resistant." This section further states, "[w]hen unattended, vehicular magazines must have wheels removed or otherwise be effectively immobilized by kingpin locking devices or other methods approved by the Director."

The regulation at 27 CFR 555.215 states, in part, "[v]olatile materials are to be kept a distance of not less than 50 feet from outdoor magazines."

The regulation at 27 CFR 555.22 allows for the approval and use of an alternate method or procedure in lieu of a method or procedure specifically prescribed in Part 555. ATF may approve an alternate method or procedure when it is found that—

(1) Good cause is shown for the use of the alternate method or procedure;

(2) The alternate method or procedure is within the purpose of, and consistent with the effect intended by, the specifically prescribed method or procedure and that the alternate method or procedure is substantially equivalent to that specifically prescribed method or procedure; and

(3) The alternate method or procedure will not be contrary to any provision of law and will not result in an increase in cost to the Government or hinder the effective administration of 27 CFR Part 555.

ATF has approved a significant number of variances for an alternate means of immobilizing preloaded bulk delivery vehicles when some additional security measures were put in place.

Bulk delivery vehicles are routinely utilized for on-site delivery of blasting services. Often these trucks contain a blasting agent as defined under 27 CFR 555.11. Utilization of these bulk products delivered on-site and used immediately has increased safety and security by reducing the number of remotely located storage trailers containing packaged blasting agent products. These delivery vehicles generally leave the explosives storage locations at unusual times of day or night, and for safety reasons they are loaded the day before for next day's delivery. These trucks are incapable of being disabled by a kingpin locking device, and the requirement to remove the wheels for immobilization is obviously not feasible. Additionally, there are times when explosive material remains on the vehicle when it returns from use. Most of these products degrade during handling therefore, removing them from the bulk vehicle would not be an option.

One of the major dangers surrounding explosives is fire. Therefore, the regulations at 27 CFR 555.215 require that volatile materials be maintained a distance of not less than 50 feet from outdoor explosive storage magazines. Because these preloaded storage vehicles contain a fuel tank filled with volatile materials, they must remain in an area protected from fire such as gravel, paved, or closely mowed designated parking area.

ATF believes that the following alternate method of operation is substantially equivalent to the prescribed methods. Increased safety for employees and the public provides good cause for this alternate method. It is not contrary to law and will not result in any increased cost to the Government. Overall, ATF believes allowing for this flexibility assists with the effective administration of 27 CFR Part 555.

Held, ATF will approve alternate methods or procedures for the preloading and temporary storage of bulk blasting agents in delivery vehicles at explosive magazine site(s), when the security and immobilization meets the following criteria:

1. All doors on the vehicle are locked, the ignition key is removed, and the key is secured away from the truck.

2. When the site is not in operation, outer perimeter security is established. This may be by a variety of means such as a locked gate, security guards, fence, natural features, or a combination of these.

3. Each potential access point to explosive materials on a storage vehicle will be secured with a minimum of one padlock that has at least five tumblers and casehardened shackle of at least ⅜" diameter.

4. Each vehicle shall be immobilized through the use of a steering wheel locking device, lockable battery disconnect switch, or both.

5. All vehicles preloaded with blasting agents shall be parked in a company designated area not susceptible to fire propagation such as bare dirt, gravel, rock, paving, or closely mowed parking lot.

All other provisions of 27 CFR Part 555 must be complied with as prescribed.

Held further, licensees and permittees who wish to use the alternate method or procedure set forth in this ruling are not required to obtain advance approval in accordance with 27 CFR 555.22.

Held further, this ATF-approved alternate method or procedure for the temporary storage of bulk blasting agent products in locked and properly secured vehicles shall not be withdrawn unless the holder of said variance is so advised by ATF in writing or no longer holds a Federal explosives license or permit.

Date approved: January 26, 2007

18. 18 U.S.C. 842(j): Storage of Explosives

18 U.S.C. 845(a): Exceptions; Relief from Disabilities

27 CFR 555.22: Alternate Methods or Procedures

27 CFR 555.29: Unlawful Storage

27 CFR 555.141: Exemptions

27 CFR 555.205: Movement of Explosive Materials

Under specific conditions, State and local bomb technicians and explosives response teams may store a limited amount of explosive materials within official response vehicles.

ATF Ruling 2009-3

The Bureau of Alcohol, Tobacco, Firearms and Explosives (ATF) has received requests from State and local law enforcement agencies to store explosive materials overnight in official response vehicles. The agencies assert that authorization for this alternate storage procedure would increase public safety and ensure that their responses to critical incidents are conducted more efficiently. Explosive storage magazines are often located several miles away from a response vehicle's location. Without a storage variance, prior to reporting to a critical incident, State and local bomb technicians and explosives response teams are often required first to travel to the explosive storage magazine location, retrieve the necessary explosive materials required to perform their duties, and then proceed to the incident scene. This situation may hinder the agency's ability to rapidly respond to critical incidents.

Under Title 18, United States Code (U.S.C.), Section 842(j), all persons must store explosive materials in compliance with the regulations issued by the Attorney General. The Attorney General has delegated the authority to administer and enforce the Federal explosives laws to the Director, ATF. The regulations contained within Title 27, Code of Federal Regulations (CFR), Part 555 implement the provisions of the Federal explosives laws. Section 555.205 provides that all explosive materials must be kept in locked magazines meeting the standards in Part 555, Subpart K, unless they are in the process of manufacture, being physically handled in the operating process of a licensee or user, being used, or being transported to a place of storage or use by a licensee or permittee or by a person who has lawfully acquired explosive materials under Section 555.106. State and local governments are not exempt from the requirement to store explosives in conformity with ATF regulations.

The regulations at 27 CFR 555.22 allow the Director, ATF, to approve the use of an alternate method or procedure in lieu of a method or procedure specifically prescribed in Part 555. ATF may approve an alternate method or procedure when: (1) Good cause is shown for the use of the alternate method or procedure; (2) The alternate method or procedure is within the purpose of, and consistent with the effect intended by, the specifically prescribed method or procedure and that the alternate method or procedure is substantially equivalent to that specifically prescribed method or procedure; and (3) The alternate method or procedure will not be contrary to any provision of law and will not result in an increase in cost to the Government or hinder the effective administration of 27 CFR Part 555.

ATF finds that, provided certain conditions are met, there is good cause for authorizing storage of explosive materials in official response vehicles. For purposes of this ruling, the term "official response vehicle" is limited to State and local law enforcement department-issued vehicles specially designated for use by bomb technicians and explosives response teams.

Allowing State and local bomb technicians and explosives response personnel the flexibility to store explosive materials overnight in official response vehicles, whether or not attended, would increase public safety and facilitate quick and efficient incident response and law enforcement operations. In addition, ATF finds that the additional security protocols provided in this ruling, such as enhanced storage security features, limited overall explosive materials load, magazine inspections, and increased inventory requirements are consistent with the effect intended by, and are substantially equivalent to, the specifically prescribed methods and procedures prescribed in Part 555, Subpart K. Further, this alternate method is not contrary to any provision of law, will not increase costs to ATF, and will not hinder the effective administration of the regulations.

Held, State and local bomb technicians and explosives response teams may store explosive materials in official response vehicles parked inside a secured building, provided the conditions set forth below are met at all times. A building is considered "secured" if it is a law enforcement or other government facility not accessible by unauthorized personnel. A secured building has law enforcement or other government personnel present at all times, or the building has an additional security feature such as an alarm, camera, or card entry system.

(1) Official response vehicles and buildings must be locked and secured at all times when not in use; and

(2) No more than 50 pounds Net Explosives Weight total may be stored in each building. This means that the combined Net Explosives Weight stored in official response vehicles and other explosives storage magazines located in the building may not exceed 50 pounds. Note that the Net Explosives Weight of explosive materials stored is not to be confused with the TNT equivalency weight of explosive materials.

Held further, State and local bomb technicians and explosives response teams may store explosive materials in unattended, official response vehicles parked at an outdoor location, provided the conditions set forth below are met at all times. The outdoor location may be an unsecured area accessible by civilians or unauthorized personnel.

(1) When not in use, official response vehicles must be locked at all times and have at least one additional security feature, such as a vehicle alarm, vehicle tracking device, or vehicle immobilization mechanism, or other equivalent alternative; and

(2) Official response vehicles located at an outdoor location may not store explosive materials in excess of:

(i) 20 detonators (electric, non-electric, or electronic); and

(ii) 2.5 pounds Net Explosives Weight of all other explosive materials.

Note that the above list does not limit the amount of disruptor ammunition or unmixed binary explosives carried on a response vehicle.

Held further, all State and local law enforcement agencies intending to store explosive materials in official response vehicles, whether attended or unattended, must meet the following criteria at all times:

(1) Explosive materials must be stored in at least a Type-3 magazine;

(2) Magazines must be secured with one steel padlock (which need not be protected by a steel hood) having at least five tumblers and a case-hardened shackle of at least ⅜" diameter. Alternatively, the magazine may be secured by placing it inside a locked compartment within the vehicle designed to meet law enforcement construction standards for weapons storage within the vehicle;

(3) Agencies must securely bolt or otherwise affix the magazines, or the locked compartments in which the magazines are stored, to the vehicle. Nuts must be located on the inside of the magazine or compartment where they cannot be removed from the outside. The nuts must be covered with a non-sparking material, such as epoxy paint or plywood;

(4) If a magazine placed in a vehicle uses a secondary locking system containing a chain or cable and a padlock, the agency need not bolt it to the trunk or cargo area of the vehicle or lock it with one steel padlock. Rather, the agency must close and stabilize the magazine securely within the trunk or cargo area of the vehicle using the secondary lock's chain or cable and padlock;

(5) Agencies may store detonators in the same magazine as delay devices, electric squibs, safety fuse, igniters, igniter cord, and shock tube, but not in the same magazine with other explosive materials;

(6) Agencies may not store any amount of loose or free-flowing explosive powders, irrespective of the packaging configuration. This does not prohibit the transport of necessary amounts of black or smokeless powders for use at specifically planned operations or the transport of seized black or smokeless powders as a result of operations;

(7) Agencies may not store any tools or other metal devices in the same magazine as the explosive materials;

(8) Officers storing explosive materials within official response vehicles must maintain an inventory storage record. The record must contain the name of the explosive material's manufacturer, the quantity on hand, and the dates that the materials are received, removed, or used. Officers must maintain a copy of this record within the vehicle and at an off-site location, such as with the Bomb Squad Commander;

(9) Officers must conduct a quarterly inventory of the explosive materials on hand and compare it to the inventory storage record. Officers must note this inventory in the inventory storage record;

(10) Officers must inspect the magazine once every 7 days to determine whether there has been any attempted or unauthorized entry into the magazine, or unauthorized removal of the contents stored in the magazine; and

(11) In the event of the theft or loss of explosive materials, law enforcement officers must report the theft or loss to ATF within 24 hours of discovery by calling 1-800-800-3855 and completing an ATF Form 5400.5, *Report of Theft or Loss of Explosive Materials*. You may obtain this form from the ATF Distribution Center by calling 301-583-4696, or through the ATF website at http://www.atf.gov/forms/download/atf-f-5400-5.pdf.

Any other alternate methods or procedures not permitted by this ruling may be requested through a variance request addressed to the ATF Explosives Industry Programs Branch (EIPB). Requests may be made by sending an email to EIPB@atf.gov or letter to:

Explosives Industry Programs Branch
Bureau of Alcohol, Tobacco, Firearms and Explosives
99 New York Avenue, NE.
Mailstop 6E403
Washington, DC 20226

To request a copy of ATF's *Federal Explosives Laws and Regulations*, ATF P 5400.7, you may contact the ATF Distribution Center at 301-583-4696, or your local ATF office. You may also download a copy of this publication from ATF's website at http://www.atf.gov/publications/explosives-arson/.

Date approved: June 19, 2009

19. 18 U.S.C. 842(j): Storage of Explosive Materials

27 CFR 555.22: Alternate Methods or Procedures

27 CFR 555.214(b): Storage within Types 1, 2, 3, and 4 Magazines

ATF authorizes an alternate method or procedure from the explosive materials visible marks storage requirement of 27 CFR 555.214(b). Specifically, ATF authorizes Federal explosives licensees and permittees to store in magazines containers of explosive materials so that marks are not visible, provided all of the requirements stated in this ruling are met.

ATF Ruling 2010–2

The Bureau of Alcohol, Tobacco, Firearms and Explosives (ATF) has received questions from Federal explosives licensees and permittees (licensees and permittees) regarding the requirements under 27 CFR 555.214(b) pertaining to the storage of containers of explosive materials in magazines.

Licensees and permittees have informed ATF that it is often impractical to store explosive materials in such a way that the label on each container of explosive materials is visible. They have advised ATF that common industry practices—developed to maximize efficiencies of labor and space—often rely on storage configurations in which the labels on some containers of explosive materials are not visible.

For example, multiple containers of explosive materials may be stacked on pallets for shipment from an importer, manufacturer, or distributor. While such a stacked-pallet arrangement of explosive materials may be beneficial for shipping purposes, the arrangement does not lend itself to easy access to and identification of every container within a stacked formation on a pallet. More specifically, explosive materials containers positioned in the center of the pallet are not readily visible and therefore the labels and marks on such containers moved into a magazine on a shipping pallet are not visible when inventory is conducted.

Similarly, magazine space limitations may compel licensees and permittees to arrange containers of explosive materials in consecutive rows, with little or no space between the rows. As a result, and similar to the cube-pallet arrangement, the labels and marks on the containers of explosive materials that are not at the front of such stacked consecutive rows are not visible for inspection.

The Federal explosives laws, at Title 18, United States Code (U.S.C.), Chapter 40, require all persons to store explosive materials in a manner in conformity with regulations issued by the Attorney General. Under 28 CFR 0.130(a), the Attorney General has delegated his authority to administer and enforce the Federal explosives laws to the Director of ATF. The regulations in 27 CFR Part 555 implement the provisions of the Federal explosives laws. The regulation at 27 CFR 555.214(b) states: "Containers of explosive materials are to be stored so that marks are visible. Stocks of explosive materials are to be stored so they can be easily counted and checked upon inspection." Marks include, among other information, the manufacturer's name and date/shift code.

Licensees and permittees may seek approval from ATF to use an alternate method or procedure in lieu of a method or procedure specifically prescribed in the regulations. Federal regulations at 27 CFR 555.22 provide that the Director of ATF may approve an alternate method or procedure, subject to stated conditions, when he finds that: (1) good cause is shown for the use of the alternate method or procedure; (2) the alternate method or procedure is substantially equivalent to, within the purpose of, and consistent with the effect intended by, the specifically prescribed method or procedure; and (3) it will not be contrary to any provision of law and will not result in an increase in cost to the Government or hinder the effective administration of 27 CFR Part 555.

ATF recognizes that, provided certain conditions are met, stocks of the same explosive materials with identical marks can be stored so that even though all marks on containers are not visible, the stocks can still be counted and checked upon inspection to readily disclose the information required for inventory and inspection. ATF has determined that explosive materials storage consistent with these practices complies with the underlying purpose of the regulation at 27 CFR 555.214(b) so long as: (1) the explosive materials containers stored in close proximity to each other are the same explosive materials with identical marks; and (2) the stocks of explosive materials can reasonably be accessed, counted, and checked upon inspection in accordance with 27 CFR 555.214(b). This storage configuration is acceptable because: (1) an accurate accounting can be conducted with minimal movement of explosive materials; and (2) randomly sampling several containers with identical marks to verify that the explosive materials are the same does not impose a significant burden on the ATF official who conducts the inventory.

A mixed storage configuration is one in which stacked containers hold different quantities, sizes and/or types of explosive materials; or hold the same quantities, sizes and types of explosive materials, but are labeled with different date/shift codes. In these configurations, the containers are often stacked so that the labels on some of the containers are hidden from view by other stacked containers. ATF has determined that such a mixed storage configuration is acceptable so long as: (1) the licensee or permittee maintains and keeps available for inspection an accurate, complete, and updated list of all the explosive materials on the pallet or in the stacked group, including the marks for each container (manufacturer's name or brand name and date/shift code), and the quantity and description (type of explosive materials); and (2) the licensee or permittee ensures that the stocks of explosive materials can reasonably be accessed, counted, and checked upon inspection in accordance with 27 CFR 555.214(b). This storage configuration is acceptable because: (1) an accurate accounting can be conducted with minimal movement of explosive materials; and (2) randomly sampling several containers to verify the accuracy of the list does not impose a significant burden on the ATF official who conducts the inventory.

The purpose of the regulation at 27 CFR 555.214(b) is to ensure that explosive materials inventories can be efficiently conducted by the licensee or permittee and by ATF officials during an inspection while minimizing the movement of explosive materials during such inventory and inspection activities. In storage scenarios like those described above, under the conditions set forth in this ruling, the fact that the marks on some containers are not immediately and readily visible does not negatively affect the ability of the ATF official or the licensee or permittee to conduct an inventory verification. ATF therefore finds there is good cause to authorize a variance from the provisions of 27 CFR 555 that require the marks on all containers of explosive materials stored in magazines to be visible. Further, this alternate method or procedure is not contrary to any provision of law, will not increase costs to ATF, and will not hinder the effective administration of the regulations.

Held, licensees and permittees may store in their magazines containers of explosive materials that have identical marks on the labels, such that the labels on some of the containers are not readily visible, so long as the stocks of explosive materials can reasonably be accessed, counted, and checked upon inspection in accordance with 27 CFR 555.214(b).

Held further, licensees and permittees may store in their magazines containers of different explosive materials, or containers of the same explosive materials with different date/shift codes on the labels, such that the labels on some of the containers are not readily visible, so long as the licensee or permittee: (1) maintains and keeps available for inspection an accurate, complete, and updated list of all the explosive materials on the pallet or in the stacked group, including the marks for each container (manufacturer's name or brand name and date/shift code), and the quantity and description (type of explosive materials); and (2) ensures that the stocks of explosive materials can reasonably be accessed, counted, and checked upon inspection in accordance with 27 CFR 555.214(b).

Date approved: June 4, 2010

20. 18 U.S.C. 842(j): Storage of Explosives

27 CFR 555.22: Alternate Methods or Procedures
27 CFR 555.205: Movement of Explosive Materials

ATF authorizes an alternate method or procedure from the provisions of 27 CFR 555.205 requiring the storage of explosive devices inside a locked magazine. Specifically, ATF authorizes explosives licensees and permittees to store loaded perforating guns outside of a locked magazine provided all of the requirements stated in this ruling are met.

ATF Ruling 2010–7

The Bureau of Alcohol, Tobacco, Firearms and Explosives (ATF) has received requests from members of the explosives industry for permission to store loaded perforating guns, which contain explosive materials, outside of a locked magazine.

Perforating guns are used primarily to pierce oil and gas wells in preparation for production operations. Industry members who use perforating guns assemble each perforating gun to meet a specific purpose and function depending on oil or gas well conditions. Each perforating gun assembly can measure up to 70 feet in length and weigh in excess of 150 pounds when loaded with explosives. Perforating gun assemblies are commonly housed within hollow thick-walled steel tubes and contain shaped charges that explode to pierce the well. Each individual shaped charge typically contains between 25 and 45 grams of high explosives and is initiated by detonating cord and a detonator. Although detonators are commonly attached to the assembly immediately prior to down-hole loading at the job site, perforating guns are sometimes armed with detonators upon assembly.

During transportation of perforating guns, Department of Transportation regulations require a detonator interrupter device to be placed between the detonators and detonating cord, which prevents the possibility of an unintended initiation.

An oil or gas well operation may urgently need perforating guns due to the varying conditions within the well operations. Assembling the guns at the worksite, often under unsafe or adverse conditions, can significantly delay operations and place workers at risk. In addition, waiting until perforating guns are needed before beginning the assembly operation would result in costly delays for the well operator. To facilitate perforating operations and meet job demands, many perforating gun operators maintain a fixed amount of preloaded perforating gun assemblies that can be used on short notice.

Preloaded perforating guns are generally assembled by affixing open-faced shaped charges and detonating cord within a steel tube carrier and then sealing the ends of the tube. Sealing the tube protects the explosives from accidental damage, discharge, or unauthorized removal. The shaped charges sealed in the carrier assembly perforate the steel tube during initiation within the well. Some perforating gun assemblies, however, have pressure-sealed shaped charges on metallic strips or wires that do not require steel tube housings.

Under Federal law, Title 18, United States Code, Section 842(j), all persons must store explosive materials, to include explosive devices, in compliance with regulations issued by the Attorney General. The Attorney General has delegated the authority to administer and enforce the Federal explosives laws to the Director, ATF. The regulations promulgated by ATF to implement the provisions of the Federal explosives laws are codified at Title 27, Code of Federal Regulations (CFR), Part 555. The regulation at 27 CFR 555.205 requires that "[a]ll explosive materials must be kept in locked magazines meeting the standards in this subpart unless they are: (a) In the process of manufacture; (b) Being physically handled in the operating process of a licensee or user; (c) Being used; or (d) Being transported to a place of storage or use by a licensee or permittee or by a person who has lawfully acquired explosive materials under § 555.106."

Under 27 CFR 555.22, ATF may approve the use of an alternate method or procedure in lieu of a method or procedure specifically prescribed in Part 555. ATF may approve an alternate method or procedure when: (1) Good cause is shown for the use of the alternate method or procedure; (2) The alternate method or procedure is within the purpose of, and consistent with the effect intended by, the specifically prescribed method or procedure and that the alternate method or procedure is substantially equivalent to that specifically prescribed method or procedure; and (3) The alternate method or procedure will not be contrary to any provision of law and will not result in an increase in cost to the Government or hinder the effective administration of Part 555.

Perforating guns are generally assembled within a building or shop that contains the tools and materials necessary to safely prepare each perforating gun. Once assembled, the length and weight of preloaded perforating guns often make it difficult or impossible to store them within an explosives magazine as prescribed under the regulations. Moreover, because many perforating gun assemblies are housed within metal tubes, these assemblies create an unnecessary sparking hazard to other explosives contained within an explosives magazine.

Further, on occasion, operators must return unused loaded perforating gun assemblies from a jobsite to their business premises due to inclement weather conditions or jobsite changes. Without approved alternate storage areas, operators would be forced to disassemble the loaded perforating guns and remove the individual explosive products to a magazine, which can significantly increase the safety risk to operators. Allowing the storage of loaded perforating guns in alternate secure areas greatly decreases the chance of an accident occurring due to unsafe or adverse conditions at the jobsite and the need for operators to handle and move loaded perforating guns repeatedly between the assembly area and the magazine. Additionally, the length and weight of loaded perforating guns can make them difficult to steal when stored outside a magazine. For these reasons, ATF has issued numerous variances to operators allowing the storage of preloaded perforating guns inside a secure building or at an outdoor location subject to certain conditions.

Allowing the storage of preloaded perforating guns in secure buildings or areas outside a magazine will increase worker safety, provided certain conditions are met, and will likely not result in increased risk of theft. For these reasons, ATF finds that there is good cause for authorizing a variance from the provisions of 27 CFR 555.205 requiring the storage of these explosive devices inside a locked magazine. ATF further finds that the additional security measures provided in this ruling as a condition of the alternate storage methods and procedures provided below, are consistent with the effect intended by, and are substantially equivalent to, the methods and procedures prescribed in Part 555, Subpart K. Further, the alternate method or procedure authorized by this ruling is not contrary to any provision of law, will not increase costs to ATF, and will not hinder the effective administration of the regulations.

Federal explosives licensees (FELs) or Federal explosives permittees (FEPs) may store perforating guns in areas outside of locked magazines, whether located indoors or outdoors, provided all of the following conditions are met at all times:

1. Loaded perforating gun assemblies armed with detonators or initiating devices must contain a detonator interrupter device.

2. A handling cap, plug, or other closure device must be installed on both ends of loaded hollow type carrier guns. Perforating guns with exposed explosive components (e.g., those that consist of sealed charges mounted on strips or wires, and that are not mounted in a carrier assembly, must be secured so as to prevent the unauthorized removal of explosives (e.g., storage within a locked room inside a building, storage within a transportation carrier or other protective housing assembly).

3. Impact wrenches or other tools that could be used to disassemble the loaded perforating guns must be kept separate and secured from the loaded perforating guns to prevent unauthorized removal of the explosive materials.

4. The government authority having jurisdiction over fire safety in the locality in which the loaded perforating guns are being stored must be notified of the location of the loaded perforating guns.

5. All buildings, areas, or vehicles containing pre-loaded perforating guns must be visually inspected at least once every three calendar days. This inspection must be sufficient to determine whether there has been unauthorized entry or attempted entry into the building, area, or vehicle, or unauthorized removal of the perforating gun assemblies from a building, area, or vehicle.

6. A daily summary of magazine transactions (27 CFR 555.127) must be maintained for each area, building, or vehicle that contains loaded perforating guns. Quantity entries may be expressed as the number of individual perforating guns stored within each separate area, building, or vehicle. Information as to the quantity and description of explosive products contained within each individual perforating gun shall be provided to any ATF officer on request.

7. The local ATF office must be notified in writing no less than three (3) Federal business days prior to utilizing an alternate storage building, area, or vehicle pursuant to this ruling. The written notification submitted to ATF must contain, at a minimum, the following information:

    a. The FEL or FEP's name;

    b. The FEL or FEP number;

    c. The location of alternate storage;

    d. The FEL or FEP's contact information, including, at a minimum, the name of a person designated as the FEL or FEP's point of contact for the notification, as well as the address, telephone and facsimile number, and, if available, email address; and

    e. Any other information concerning the alternate storage or items stored that ATF may require.

8. The licensee or permittee desiring to store perforating guns in areas outside of magazines subject to the conditions of this ruling should maintain a copy of its submission to ATF of the information required (Item #7, above) with its permanent records. The licensee or permittee should also retain proof of its submission (e.g., certified return receipt mail, tracking receipt, or other printed verification).

9. All other provisions of 27 CFR Part 555 must be complied with as prescribed.

If the loaded perforating guns are stored in an indoor location, the following additional conditions must be met at all times:

1. No more than 50 pounds of explosives may be stored inside any facility used for indoor explosives storage. This means that the combined net explosives weight contained in magazines and outside of magazines in any one building must not exceed 50 pounds.

2. All loaded perforating guns must either lie flat on the floor or be placed in stable racks to prevent accidental movement or discharge.

3. The building in which the loaded perforating guns are stored must be securely locked or attended by FEL/FEP responsible persons or employee possessors.

If the loaded perforating guns are stored in an outdoor location, the following additional conditions must be met at all times:

1. The perimeter of the area containing loaded perforating guns must be secured by a security fence with a locked gate or the entire facility must be completely enclosed by a security fence with each entrance point secured at all times by a locked gate. For the purpose of this ruling, the security fence must be a minimum of 6 feet high and have firmly anchored posts to ensure its structural stability. The gate must be secured with a padlock that has at least five tumblers and a case-hardened or boron alloy shackle of at least ⅜" diameter.

2. Loaded perforating guns may be stored on a vehicle or trailer provided all doors are locked. The vehicle or trailer must be immobilized by a standard kingpin locking device, steering wheel locking device, or lockable battery disconnect. The ignition key must be removed from the vehicle and secured away from the vehicle.

3. All vehicles loaded with perforating guns must be parked in an area not susceptible to fire propagation such as bare dirt, gravel, rock, paving, or closely mowed grass.

4. No more than 200 loaded perforating guns or 2,500 pounds of net explosive weight may be kept outside a magazine. Each outdoor area or vehicle containing loaded perforating guns must comply with the table of distances requirements set forth in 27 CFR 555.218.

5. The loaded perforating guns may be stored in a covered or uncovered area as long as the guns are placed flat on a concrete or paved floor, offshore well tool pallets, or placed on permanently mounted racks to prevent accidental movement or discharge.

6. Loaded perforating guns that have exposed explosives that are not contained within a secure tube (e.g., tube-less strips or wire guns) may not be stored in an outdoor location.

Once the licensee or permittee has submitted the necessary documentation to ATF pursuant to this ruling, and complied with all other conditions set forth in this ruling, no separate, individual variance approval from ATF is required, and the licensee or permittee may store perforating guns in areas outside of magazines. Licensees and permittees must still abide by all other provisions relating to the storage of explosives.

Held, pursuant to 27 CFR 555.22, ATF authorizes an alternate method or procedure from the provisions of 27 CFR 555.205 requiring the storage of explosive devices inside a locked magazine. Specifically, ATF authorizes explosives licensees and permittees to store loaded perforating guns outside of a locked magazine provided all of the requirements stated in this ruling have been met.

Held further, if ATF finds that a licensee or permittee has failed to abide by the conditions of this ruling, or uses any procedure that hinders the effective administration of the explosives laws or regulations, ATF may notify the licensee or permittee that the licensee or permittee is no longer authorized to store perforating guns outside of a locked magazine under this ruling.

Date approved: November 24, 2010

21. 18 U.S.C. 842(j): Storage of Explosives

27 CFR 555.22: Alternate Methods or Procedures

27 CFR 555.211(a)(4): Locking Requirements of Type 5 Magazines

ATF authorizes an alternate method or procedure from the provisions of 27 CFR 555.211(a)(4) requiring the storage of explosives in outdoor type 5 magazines with doors equipped with the requisite locks. Specifically, ATF authorizes explosives licensees and permittees to store bulk blasting agents in outdoor type 5 bin or silo magazines without the specified locks provided all of the requirements stated in this ruling have been met.

ATF Ruling 2011–2

The Bureau of Alcohol, Tobacco, Firearms and Explosives (ATF) has received numerous requests from Federal explosives licensees and permittees for variances from the locking requirements for outdoor type 5 bin and silo storage magazines.

Outdoor type 5 bin and silo magazines are used to store bulk blasting agents, typically bulk emulsion and ammonium nitrate-fuel oil (ANFO). Due to the nature of these bulk explosive materials, they cannot be stored in magazines with conventional doors that are easily padlocked and hooded. Instead, bulk blasting agents must be stored in magazines with filling and discharge systems that have small openings that are secured and controlled by hatches, valves, and nozzles. It is impractical to retrofit the locks used on these magazines with hoods because of the configuration of the filling and discharge systems. Additionally, access points, including ladders and hatch openings, cannot be hooded or secured with two padlocks without impeding access to the magazine.

Outdoor type 5 bin and silo magazines are also inherently less susceptible to theft than conventional magazines due to their design. Bin and silo access points are small, and entry to the top of the magazine can only be gained using a pull-down ladder. The discharge valve is typically located on the underside of the bin, making it difficult to access. Additionally, the nature of the bulk blasting agents stored in these types of magazines makes them more difficult to steal than packaged explosives products. They are difficult to remove from the bins or silos due to the height of a common silo and the lack of packaging for bulk products. For these reasons, pursuant to an ATF—approved variance, licensees and permittees generally use alternate locking configurations on their type 5 bin or silo magazines, subject to certain conditions.

Under Federal law, Title 18, United States Code, section 842(j), all persons must store explosive materials in compliance with regulations issued by the Attorney General. The Attorney General has delegated the authority to administer and enforce the Federal explosives laws to the Director, ATF. The regulations promulgated by ATF to implement the provisions of the Federal explosives laws are codified at Title 27, Code of Federal Regulations (CFR), Part 555.

The regulation at 27 CFR 555.211(a)(4) addresses the locking requirements of outdoor type 5 magazines, providing that "[e]ach door is to be equipped with (i) two mortise locks; (ii) two padlocks fastened in separate hasps and staples; (iii) a combination of a mortise lock and a padlock; (iv) a mortise lock that requires two keys to open; or (v) a three-point lock. Padlocks must have at least five tumblers and a case-hardened shackle of at least ⅜" diameter. Padlocks must be protected with not less than ¼" steel hoods constructed so as to prevent sawing or lever action on the locks, hasps, and staples."

Licensees and permittees may seek approval from ATF to use an alternate method or procedure in lieu of a method or procedure specifically prescribed in the regulations. Federal regulations at 27 CFR 555.22 provide that the Director of ATF may approve an alternate method or procedure, subject to stated conditions, when he finds that: (1) good cause is shown for the use of the alternate method or procedure; (2) the alternate method or procedure is substantially equivalent to, within the purpose of, and consistent with the effect intended by, the specifically prescribed method or procedure; and (3) it will not be contrary to any provision of law and will not result in an increase in cost to the Government or hinder the effective administration of 27 CFR Part 555.

ATF finds that there is good cause for licensees and permittees to store bulk blasting agents in outdoor type 5 bin or silo magazines with doors not equipped with the requisite locks. Due to the nature of bulk blasting agents, it is difficult for licensees and permittees to comply with the locking requirements of section 555.211(a)(4). Further, the design of outdoor type 5 bin and silo magazines, and the inaccessibility of the bulk blasting agents stored within them, provides protection from theft and loss that is substantially equivalent to the requirements of section 555.211 (a)(4), provided certain conditions are met. This alternate method or procedure is also not contrary to any provision of law, will not increase costs to ATF, and will not hinder the effective administration of the regulations.

Federal Explosives Licensees (FELs) or Federal Explosives Permittees (FEPs) may store bulk blasting agents in outdoor type 5 bin or silo magazines with doors not equipped with the requisite locks, provided all of the following conditions are met at all times:

1. When not in operation, adequate security that restricts outer perimeter access must be provided. The use of a locked gate, security guards, fences, or a combination of these measures may be used.

2. Each outdoor bin or silo must be secured with at least one five tumbler padlock with a case-hardened shackle of at least ³⁄₈" diameter on the top hatch, the valves at the loading ports and points of discharge, and the elevated ladder to prevent anyone from climbing onto bins or silos.

3. Blasting agent storage silos and bins must meet all other construction requirements of type 5 magazines as described in 27 CFR 555.211.

All other provisions of 27 CFR Part 555 must be complied with as prescribed.

Once the licensee or permittee has complied with all conditions set forth in this ruling, no separate, individual variance approval from ATF is required, and the licensee or permittee may store bulk blasting agents in outdoor type 5 bin or silo magazines with doors not equipped with the requisite locks.

Held, pursuant to 27 CFR 555.22, ATF authorizes an alternate method or procedure from the provisions of 27 CFR 555.211(a)(4) requiring the storage of explosives in outdoor type 5 magazines with doors equipped with the requisite locks. Specifically, ATF authorizes explosives licensees and permittees to store bulk blasting agents in outdoor type 5 bin or silo magazines without the specified locks provided all of the requirements stated in this ruling have been met.

Held further, if ATF finds that a licensee or permittee has failed to abide by the conditions of this ruling, or uses any procedure that hinders the effective administration of the explosives laws or regulations, ATF may notify the licensee or permittee that the licensee or permittee is no longer authorized to utilize this alternate method or procedure.

Date approved: February 18, 2011

22. 18 U.S.C. 842(j): Storage of Explosives

27 CFR 555.22: Alternate Methods or Procedures

27 CFR 555.207: Construction of Type 1 Magazines

27 CFR 555.208: Construction of Type 2 Magazines

27 CFR 555.209: Construction of Type 3 Magazines

27 CFR 555.210: Construction of Type 4 Magazines

27 CFR 555.211: Construction of Type 5 Magazines

ATF authorizes an alternate method or procedure from the provisions of 27 CFR Part 555, Subpart K, requiring the storage of explosives in magazines with doors equipped with the requisite locks. Specifically, ATF authorizes explosives licensees and permittees to secure explosives magazines with hidden-shackle "hockey puck" locks, recessed padlocks, and padlocks with boron alloy shackles, provided all of the requirements stated in this ruling have been met.

ATF Ruling 2011–3

The Bureau of Alcohol, Tobacco, Firearms and Explosives (ATF) has received requests from members of the explosives industry for permission to secure explosives magazines with certain alternate locks or locking methods.

Recent advances in locking technologies have provided the explosives industry with more options for securing their explosives storage magazines. Currently, explosives industry members secure their explosives magazines primarily with padlocks fastened in separate hasps and staples and protected with ¼" thick steel hoods. However, the commercial availability of steel hoods is limited to a small number of distributors, and non-commercial fabrication can be costly.

As a more cost effective alternative, explosives industry members have asked ATF to approve hidden-shackle locks, commonly known as "hockey puck" or "puck" locks to secure their explosives storage magazines. Several companies manufacture hidden-shackle puck locks and protective steel hasps that are distributed by local hardware and construction retail outlets throughout the United States. Hidden-shackle puck locks are generally based on the same design, in which the lock shackles are completely enclosed within the steel lock body, thereby protecting the shackle from cutting or sawing action. The lock cylinders and other internal mechanisms within the hidden-shackle puck locks are designed similarly to those in traditional padlocks. The locks are commonly sold with steel hasps that are affixed with steel staples and with steel shrouds that protect the locks and staples from prying or lever action. Most steel hasps are installed with ³⁄₈" thick carriage bolts.

Explosives industry members have also asked ATF to approve magazines such as job site boxes, typically used at construction worksites and other industrial locations, as type 4 magazines for the storage of low explosives. Several companies manufacture job site boxes that are available from local hardware and construction retail outlets throughout the United States. Generally, job site boxes are based on a design that accepts padlocks housed into recessed openings that prevent the sawing, levering, or prying action on the locks, staples, or hasps. The boxes are typically constructed of 12 or 16 gauge steel and affixed with top-opening doors (lids) that lock when two internal metal latches are secured by two recessed padlocks. The padlocks are inserted shackle first into an opening of a shape and size that fully accepts the lock, and the lock shackles are secured to metal u-bolt staples. The padlocks and u-bolts cannot be removed without gaining access to the magazine interior. The padlock base and keyway are the only visible portions of the padlocks. Job site boxes are not manufactured with ¼" steel hoods to protect the padlocks, though the magazine body acts as a protective steel hood for the recessed padlocks.

Additionally, explosives industry members have asked ATF to approve padlocks that contain boron alloy steel shackles in lieu of case-hardened shackles. The manufacture of case-hardened shackles involves hardening the surface of low carbon steel by infusing carbon into the steel's surface mainly through flame or induction hardening. Case-hardening is usually conducted after the steel shackle has been formed into its final shape and its center remains tough and malleable. Boron has been found as a useful alloy to improve the hardness of steel. Manufacturing boron alloy shackles involves hardening the entire shackle by heat-treating medium carbon steel thereby creating a consistent hardness throughout the steel. Common testing methods used in the manufacturing industry to test the strength of different materials reveal that boron alloy steel is stronger than case-hardened steel.

Under Federal law, Title 18, United States Code, section 842(j), all persons must store explosive materials in compliance with regulations issued by the Attorney General. The Attorney General has delegated the authority to administer and enforce the Federal explosives laws to the Director, ATF. The regulations promulgated by ATF to implement the provisions of the Federal explosives laws are codified at Title 27, Code of Federal Regulations (CFR), Part 555.

The regulations at 27 CFR 555.207, 555.208, 555.210, and 555.211 address the locking requirements for types 1, 2, 4, and 5 explosives magazines, providing generally that each door is to be equipped with (i) two mortise locks; (ii) two padlocks fastened in separate hasps and staples; (iii) a combination of a mortise lock and a padlock; (iv) a mortise lock that requires two keys to open; or (v) a three-point lock. Padlocks must have at least five tumblers and a case-hardened shackle of at least ⅜" diameter. Padlocks must be protected with not less than ¼" steel hoods constructed so as to prevent sawing or lever action on the locks, hasps, and staples. The regulation at 27 CFR 555.209 addresses the locking requirements for type 3 explosives magazines,

providing that doors are to be equipped with one steel padlock (which need not be protected by a steel hood) having at least five tumblers and a case-hardened shackle of at least ⅜" diameter.

Licensees and permittees may seek approval from ATF to use an alternate method or procedure in lieu of a method or procedure specifically prescribed in the regulations. Federal regulations at 27 CFR 555.22 provide that the Director of ATF may approve an alternate method or procedure, subject to stated conditions, when he finds that: (1) good cause is shown for the use of the alternate method or procedure; (2) the alternate method or procedure is substantially equivalent to, within the purpose of, and consistent with the effect intended by, the specifically prescribed method or procedure; and (3) it will not be contrary to any provision of law and will not result in an increase in cost to the Government or hinder the effective administration of 27 CFR Part 555.

Historically, the most prevalent method of unauthorized entry into explosives magazines has been through attacks against the locking mechanisms—specifically against the lock shackles. Hidden shackle "hockey puck" locks, recessed padlocks, and padlocks with boron alloy shackles provide shackle protection equivalent to traditional case-hardened padlocks and ¼" steel hoods. Allowing storage of explosives in magazines secured with these types of locks will also reduce costs for explosives licensees and permittees and give them greater flexibility to secure their explosives. For these reasons, ATF finds that, provided certain conditions are met, there is good cause for licensees and permittees to store explosives in magazines with doors equipped with hidden shackle "hockey puck" locks, recessed padlocks, and padlocks with boron alloy shackles. ATF further finds that the conditions provided in this ruling for the alternate storage methods and procedures provided below, are consistent with the effect intended by, and are substantially equivalent to, the methods and procedures prescribed in Part 555, Subpart K. Further, the alternate methods or procedures authorized by this ruling are not contrary to any provision of law, will not increase costs to ATF, and will not hinder the effective administration of the regulations.

Federal explosives licensees and permittees may store explosives in magazines that are secured with hidden-shackle "hockey puck" locks, whether located indoors or outdoors, provided all of the following conditions are met at all times:

1. Each magazine door must be secured with the same number of hidden-shackle puck locks as prescribed in the regulations (e.g. two puck locks on outdoor type 4 magazines, one puck lock on mobile type 3 and type 5 magazines).

2. The hidden shackle puck lock body must be constructed of hardened steel and contain at least a five-pin tumbler cylinder. The lock shackle must be constructed of a case-hardened steel or boron alloy and measure a minimum nominal diameter of ⅜".

3. Each hidden-shackle puck lock must be protected within a solid steel hasp and shroud, or by a ¼" thick steel hood that prevents the prying or lever action on the puck lock.

4. The steel hasp must contain a ¼" thick steel shroud that surrounds the lock. Openings in the shroud required to access the lock keyway and open the magazine door must be small enough to prevent sawing, levering, or prying action on the puck lock.

5. The spaces between the steel hasps and locks, and steel shrouds and locks, must be small enough to prevent sawing, levering, or prying action on the puck lock.

6. The hasp or hood must be attached to the magazine doors by welding, or installed with at least ⅜" thick carriage bolts (with nuts on inside of door) so that they cannot be removed when the doors are closed and locked.

Federal explosives licensees and permittees may store explosives in magazines that are secured with recessed padlocks, whether located indoors or outdoors, provided all of the following conditions are met at all times:

1. Each magazine door (lid) must be secured with two recessed padlocks that have at least five tumblers and a case-hardened steel or boron alloy shackle of at least ⅜" diameter.

2. The recessed opening that houses the locks must be small enough to prevent sawing, levering, or prying action on the locks when the locks are installed.

3. The lock shackles must be securely affixed to the interior staples so the padlocks cannot be removed without gaining access to the magazine interior.

Once the licensee or permittee has complied with all conditions set forth in this ruling, no separate, individual variance approval from ATF is required, and the licensee or permittee may store explosives in magazines with doors equipped with locks other than those required by 27 CFR Part 555. Licensees and permittees are reminded of their responsibility to abide by all other provisions of 27 CFR Part 555 as prescribed.

Held, pursuant to 27 CFR 555.22, ATF authorizes an alternate method or procedure from the provisions of 27 CFR Part 555, Subpart K, requiring the storage of explosives in magazines with doors equipped with the requisite locks. Specifically, ATF authorizes explosives licensees and permittees to secure explosives magazines with hidden-shackle "hockey puck" locks, recessed padlocks, and padlocks with boron alloy shackles, provided all of the requirements stated in this ruling have been met.

Held further, if ATF finds that a licensee or permittee has failed to abide by the conditions of this ruling, or uses any procedure that hinders the effective administration of the explosives laws or regulations, ATF may notify the licensee or permittee that the licensee or permittee is no longer authorized to utilize this alternate method or procedure.

Held further, this ruling supersedes all previous variance approvals for explosives magazines secured with hidden-shackle "hockey puck" locks or recessed padlocks.

Date approved: June 23, 2011

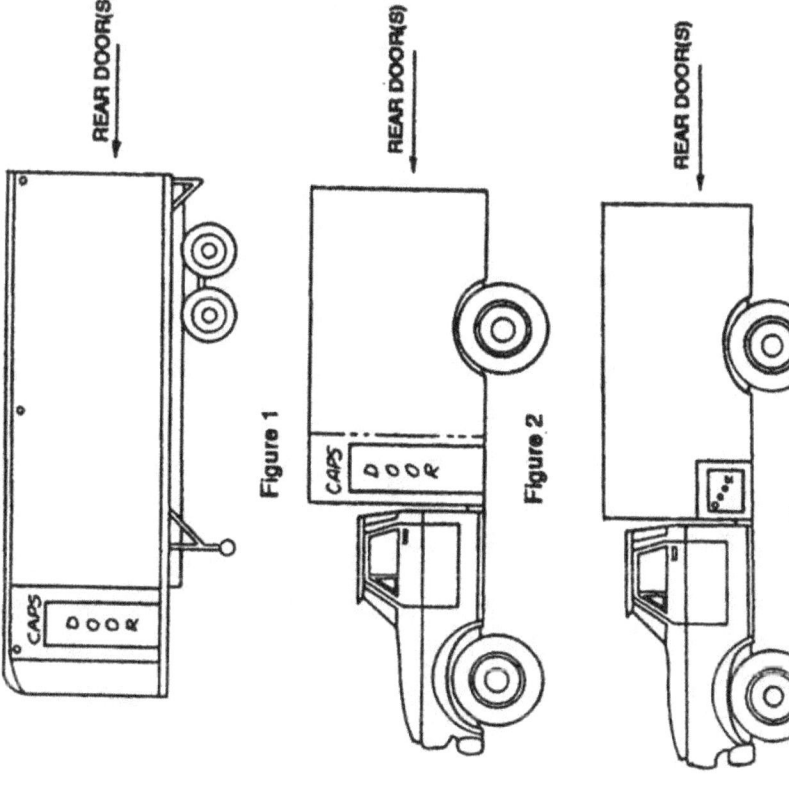

Figure 1

Figure 2

REAR DOOR(S)

REAR DOOR(S)

REAR DOOR(S)

Figure 3

NOTE: The configurations shown in Figures 1 and 2 are equally applicable to multi-axle and "cab-over" vehicles.

[Diagrams: Courtesy of IME]

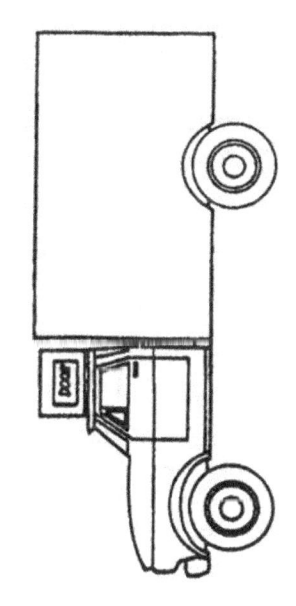

Figure 1

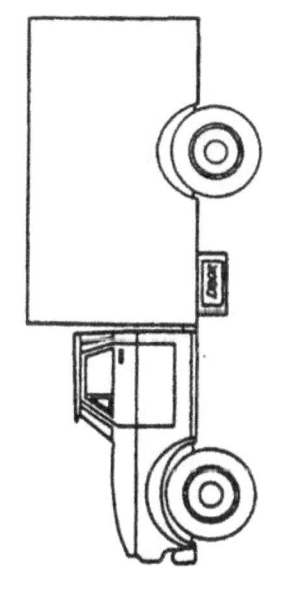

Figure 2

NOTE: The configurations shown in Figures 1 and 2 are equally applicable to multi-axle and "cab-over" vehicles.

[Diagrams: Courtesy of IME]

ATF Rul. 77-24, Appendix A and B

APPENDIX E

PORTABLE WHEELED TRAILERS

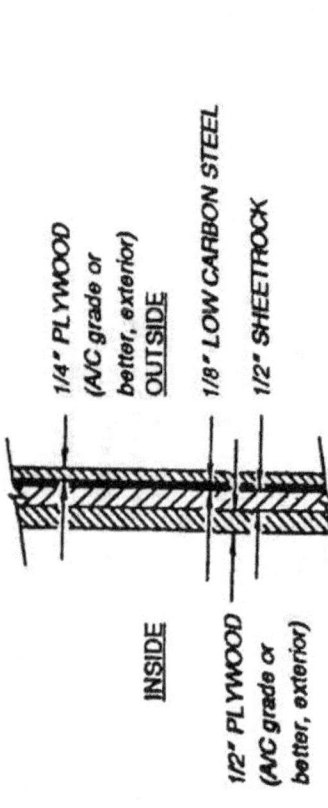

Cap Storage

Figure 1

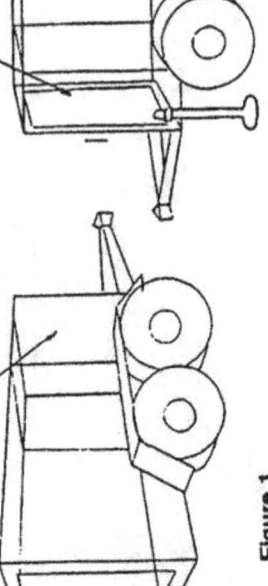

Cap Storage

Figure 3

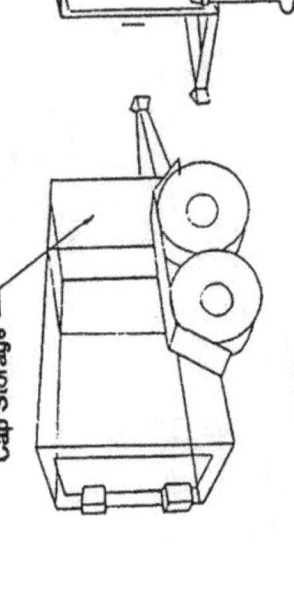

Cap Storage
Cap Storage

Figure 2

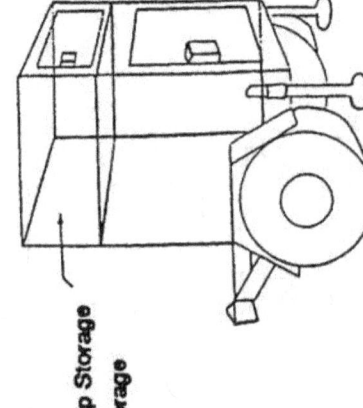

Cap Storage

Figure 4

APPENDIX C

INSIDE

1/4" PLYWOOD
(A/C grade or
better, exterior)
OUTSIDE

1/8" LOW CARBON STEEL

1/2" SHEETROCK

1/2" PLYWOOD
(A/C grade or
better, exterior)

Sketch of laminate construction for container or compartment for electric blasting caps
use, as illustrated in Appendix A, B, and E.

APPENDIX D

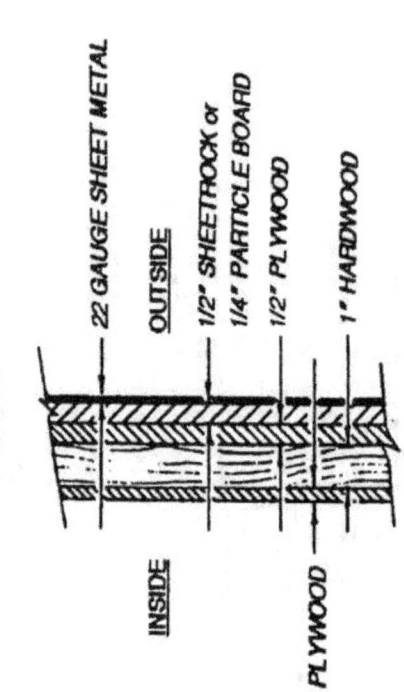

INSIDE

22 GAUGE SHEET METAL

OUTSIDE

1/2" SHEETROCK or
1/4" PARTICLE BOARD
1/2" PLYWOOD

1" HARDWOOD

1/4" PLYWOOD

Sketch of laminate construction for container or compartment for electric blasting caps;
restricted to use as illustrated in Appendix A.

[Diagrams: Courtesy of IME]

ATF Rul. 77–24, Appendix C – E

Effect of 18 U.S.C. Chapter 40 On the Fireworks Industry

[Caution! This item discusses Federal requirements only. Please contact your State or local authorities for any additional requirements.]

Title XI of the Organized Crime Control Act of 1970 (18 U.S.C. Chapter 40) establishes controls over explosive materials, including black powder and other pyrotechnic compositions commonly used in fireworks. Part 555 of Title 27, Code of Federal Regulations (CFR), contains the regulations which implement Title XI. Section 555.141(a)(7) exempts "the importation, distribution, and storage of fireworks classified as UN0336, UN0337, UN0431, or UN0432 explosives by the U.S. Department of Transportation at 49 CFR 172.101 and generally known as 'consumer fireworks' or 'articles pyrotechnic.'" Section 555.141(a)(7) does not exempt "display fireworks," as defined in 555.11.

With Respect to Fireworks: Who needs a license?

1. Manufacturers of black powder;
2. Manufacturers of any other explosive material used in manufacturing consumer fireworks or display fireworks; and
3. Importers of, or dealers in, display fireworks.

With Respect to Fireworks: Who needs a permit?

1. All persons transporting, shipping, causing to be transported, or receiving display fireworks, regardless of whether for their own use or for commercial display purposes (Certain exemptions apply, e.g. agencies of the United States or of any State or political subdivisions thereof are exempt from permit requirements); and
2. A person, other than a licensee, transporting, shipping, causing to be transported, or receiving explosive materials for use in manufacturing display fireworks or consumer fireworks.

With Respect to Fireworks: Who may not need a license or permit?

Frequently, persons contracting for display fireworks (e.g., for Fourth of July observances) from a Federal explosives licensee or permittee receive a total service, including the services of a pyrotechnician who transports display fireworks in interstate or intrastate commerce to the site of the display and conducts and supervises the display. In these instances, the customers purchase and receive the contractor's services and not the explosive materials themselves (i.e. the cost of the services includes the contractor's expense in providing the fireworks utilized), and the cost of the services includes the dealer's expense in providing the fireworks utilized. When business is transacted in this manner, the customers purchasing and receiving the services need not obtain Federal explosives licenses or permits under Part 555 as long as they are not transporting, shipping, causing to be transported, or receiving explosive materials. Note: the transportation of explosive materials to the display sites would be authorized by the Federal explosives license or permit of the licensee or permittee providing the services.

With Respect to Fireworks: Types of permits

1. User permit: Allows the permit holder to transport, ship, cause to be transported, and receive display fireworks in interstate or foreign commerce for his or her own use and not for resale. This permit is issued at a cost of $100 for a 3-year period and is renewable at a cost of $50 for a 3-year period.
2. User-limited permit: Identical to the user permit but issued for a single purchase transaction, only. The fee is $75; the permit is nonrenewable.

With Respect to Fireworks: Storage

The law prohibits any person from storing any explosive materials in a manner not in conformity with the regulations promulgated by the Attorney General (18 U.S.C. 842(j)). Pursuant to this section, the Attorney General has prescribed storage regulations in 27 CFR Part 555, Subpart K. Display fireworks must be stored in conformity with the regulations. Display fireworks generally contain perchlorate mixture explosives, potassium chlorate base explosive mixtures, and black powder, which are entered on the List of Explosive Materials with numerous others. (The List, which is not all-inclusive, is annually compiled and readily available without charge from the address set out in 27 CFR 555.23 or online at www.atf.gov/.) Display fireworks must be stored as low explosives in magazines meeting, at a minimum, the requirements for type 4 storage magazines prescribed by 27 CFR 555.210 unless they contain other classes of explosives. Bulk salutes must be stored as high explosives in type 1 or type 2 magazines. The net weight of the explosive materials contained in the display fireworks may be used in determining compliance with table of distance requirements. To determine the actual weight of the materials, it may be necessary to contact their manufacturers. The manufacturer of exempt or nonexempt fireworks having stocks of explosive materials on hand to be used in the manufacture of fireworks must store the stocks in conformity with applicable storage requirements.

Explosives Dealer's and User's Guide to Federal Explosives Regulation

Explosives May Not Be Distributed by Licensees (Or by Any Person) to Any Person Who:

1. Is under indictment for, or who has been convicted of a crime punishable by imprisonment for a term exceeding one year.

2. Is an unlawful user of, or addicted to, marijuana or any depressant or stimulant drug or narcotic drug (as these terms are defined in section 102 of the Controlled Substances Act).

3. Has been adjudicated as a mental defective or has been committed to a mental institution.

4. Is a fugitive from justice.

5. Is an alien (with certain exceptions).

6. Has been discharged from the armed forces under dishonorable conditions; or

7. Having been a citizen of the United States, has renounced citizenship.

8. Is less than 21 years of age.

Dealers in Explosives Must:

Have a current and valid Federal explosives license.

Have proper storage facilities.

Keep accurate and complete records.

Verify that each buyer has a Federal explosives license or permit.

Verify buyers' identities.

Users of Explosives

Federal permits are required of those who transport, ship, cause to be transported, or receive explosive materials. Among other things, the permittee must keep complete and accurate records of the acquisitions and dispositions of explosives materials. Unless otherwise exempted by law, no person may receive or transport any explosive materials without a permit.

No person shall store any explosive material in a matter not in conformity with applicable regulations.

All persons must report to ATF and local authorities any loss or theft of their explosive materials within 24 hours of discovery.

A Federal license or permit does not confer any right or privilege to violate any state law or local ordinance.

The above summary is general and does not purport to fully convey the Federal explosives law and regulations pertaining to dealers and users.

Black Powder Transactions

Public Law 93-639 (1975) allows nonlicensees/nonpermittees to purchase commercially manufactured black powder, in quantities of 50 pounds or less, solely for sporting, recreational or cultural purposes for use in antique firearms or antique devices. A nonlicensee or nonpermittee purchasing black powder under the exemption need not be a resident of the State in which the dealer is located. Also, the categories of persons to whom the distribution of explosive materials is prohibited do not apply to black powder transactions made under the exemption. Acquisitions of black powder not qualifying under this exemption are subject to the same regulatory requirements that govern any other low explosive.

All persons who distribute black powder, regardless of quantity, must be licensed as explosives dealers and, among other things, must provide adequate storage.

Explosives Security

Through prompt reporting of losses and thefts of explosives and increased emphasis on physical security, explosives licensees and permittees can contribute greatly to efforts by Federal, State and local authorities to reduce the incidence of bombings and other criminal misuse of explosives in the United States. The following actions are of prime importance and in some instances required:

Report... any thefts or losses of explosives within 24 hours of discovery, by telephone, to ATF (toll free: 1-800-800-3855) and to appropriate local authorities. Because the States and many municipalities have designated specific agencies to investigate the theft or loss of explosives, licensees and permittees are urged to be familiar with State and local reporting procedures and appropriate contact points.

Follow... telephone notification with a written report on ATF Form 5400.5, *"Report of Theft or Loss— Explosive Materials,"* to the nearest ATF Division Office, and in accordance with the instructions on the form.

Observe... activity around magazines, within business premises, and on job sites, particularly if strangers appear to be loitering in the area in which explosives are being kept. On-site users should take special care to assure that explosives removed from storage for use on the job are either detonated or accounted for and unused items returned to storage.

Review... recordkeeping practices to assure that no discrepancies exist and that no figures in reported inventories have been manipulated, and correct any clerical errors promptly. Should any questions arise concerning explosives security procedures or any aspect of explosives regulation coming under the jurisdiction of ATF, do not hesitate to contact ATF.

Note: For Q&A's on regulatory requirements governing recordkeeping and storage, see "Questions and Answers" numbers 62–87.

Additional Information

The flow of useful information is an essential ingredient in the effective administration of regulatory programs. The Bureau of Alcohol, Tobacco, Firearms and Explosives is the Federal agency charged with the responsibility of administering laws impacting the explosives industry. We call your attention to the following publication distributed by ATF that merits your attention:

The Explosives Newsletter

During 1989 ATF developed the Explosives Industry Newsletter, an information service for Federal explosives licensees and permittees which is intended to help explosives industry members better understand the Federal laws under which they must operate. It also includes other items of particular interest to the explosives industry. There is no charge for the Explosives Industry Newsletter.

Explosives industry members who have questions concerning the Federal explosives laws and regulations may address their inquiries to:

> Bureau of Alcohol, Tobacco, Firearms and Explosives
> Explosives Industry Programs Branch
> 99 New York Avenue, N.E.
> Mailstop 6N672
> Washington, DC 20226

Direct e-mail inquiries on general questions or variance requests may be sent to the branch at EIPB@atf.gov

List of Explosive Materials

Pursuant to the provisions of section 841(d) of title 18, U.S.C., and 27 CFR 555.23, the Director, Bureau of Alcohol, Tobacco, Firearms and Explosives, must revise and publish in the Federal Register at least annually a list of explosives determined to be within the coverage of 18 U.S.C. Chapter 40, *Importation, Manufacture, Distribution and Storage of Explosive Materials*. The list chapter covers not only explosives, but also blasting agents and detonators, all of which are defined as explosive materials in section 841(c) of title 18, U.S.C. Accordingly, the following is the current List of Explosive Materials subject to regulation under 18 U.S.C. Chapter 40. Materials constituting blasting agents are marked by an asterisk. While the list is comprehensive, it is not all-inclusive. The fact that an explosive material may not be on the list does not mean that it is not within the coverage definitions in section 841 of title 18, U.S.C. Explosive materials are listed alphabetically by their common names, followed by chemical names and synonyms in brackets. This revised list is effective as of October 19, 2011.

List of Explosive Materials

A
Acetylides of heavy metals.
Aluminum containing polymeric propellant.
Aluminum ophorite explosive.
Amatex.
Amatol.
Ammonal.
Ammonium nitrate explosive mixtures (cap sensitive).
* Ammonium nitrate explosive mixtures (non-cap sensitive).
Ammonium perchlorate having particle size less than 15 microns.
Ammonium perchlorate explosive mixtures (excluding ammonium perchlorate composite propellant (APCP)).
Ammonium picrate [picrate of ammonia, Explosive D].
Ammonium salt lattice with isomorphously substituted inorganic salts.
* ANFO [ammonium nitrate-fuel oil].
Aromatic nitro-compound explosive mixtures.
Azide explosives.

B
Baranol.
Baratol.
BEAF [1, 2-bis (2, 2-difluoro-2-nitroacetoxyethane)].
Black powder.
Black powder based explosive mixtures.
* Blasting agents, nitro-carbo-nitrates, including non-cap sensitive slurry and water gel explosives.
Blasting caps.
Blasting gelatin.
Blasting powder.
BTNEC [bis (trinitroethyl) carbonate].
BTNEN [bis (trinitroethyl) nitramine].
BTTN [1,2,4 butanetriol trinitrate].
Bulk salutes.
Butyl tetryl.

C
Calcium nitrate explosive mixture.
Cellulose hexanitrate explosive mixture.
Chlorate explosive mixtures.
Composition A and variations.
Composition B and variations.
Composition C and variations.
Copper acetylide.
Cyanuric triazide.
Cyclonite [RDX].
Cyclotetramethylenetetranitramine [HMX].
Cyclotol.
Cyclotrimethylenetrinitramine [RDX].

D
DATB [diaminotrinitrobenzene].
DDNP [diazodinitrophenol].
DEGDN [diethyleneglycol dinitrate].
Detonating cord.
Detonators.
Dimethylol dimethyl methane dinitrate composition.
Dinitroethyleneurea.
Dinitroglycerine [glycerol dinitrate].
Dinitrophenol.
Dinitrophenolates.
Dinitrophenyl hydrazine.
Dinitroresorcinol.
Dinitrotoluene-sodium nitrate explosive mixtures.
DIPAM [dipicramide; diaminohexanitrobiphenyl].
Dipicryl sulfone.
Dipicrylamine.
Display fireworks.
DNPA [2,2-dinitropropyl acrylate].
DNPD [dinitropentano nitrile].
Dynamite.

E
EDDN [ethylene diamine dinitrate].
EDNA [ethylenedinitramine].
Ednatol.
EDNP [ethyl 4,4-dinitropentanoate].
EGDN [ethylene glycol dinitrate].
Erythritol tetranitrate explosives.
Esters of nitro-substituted alcohols.
Ethyl-tetryl.
Explosive conitrates.
Explosive gelatins.
Explosive liquids.
Explosive mixtures containing oxygenreleasing inorganic salts and hydrocarbons.
Explosive mixtures containing oxygenreleasing inorganic salts and nitro bodies.
Explosive mixtures containing oxygenreleasing inorganic salts and water insoluble fuels.

Explosive mixtures containing oxygenreleasing inorganic salts and water soluble fuels.
Explosive mixtures containing sensitized nitromethane.
Explosive mixtures containing tetranitromethane (nitroform).
Explosive nitro compounds of aromatic hydrocarbons.
Explosive organic nitrate mixtures.
Explosive powders.

F
Flash powder.
Fulminate of mercury.
Fulminate of silver.
Fulminating gold.
Fulminating mercury.
Fulminating platinum.
Fulminating silver.

G
Gelatinized nitrocellulose.
Gem-dinitro aliphatic explosive mixtures.
Guanyl nitrosamino guanyl tetrazene.
Guanyl nitrosamino guanylidene hydrazine.
Guncotton.

H
Heavy metal azides.
Hexanite.
Hexanitrodiphenylamine.
Hexanitrostilbene.
Hexogen [RDX].
Hexogene or octogene and a nitrated Nmethylaniline.
Hexolites.
HMTD [hexamethylenetriperoxidediamine].
HMX [cyclo-1,3,5,7-tetramethylene 2,4,6,8-tetranitramine; Octogen].
Hydrazinium nitrate/hydrazine/aluminum explosive system.
Hydrazoic acid.

I
Igniter cord.
Igniters.
Initiating tube systems.

K
KDNBF [potassium dinitrobenzo-furoxane].

L
Lead azide.
Lead mannite.
Lead mononitroresorcinate.
Lead picrate.
Lead salts, explosive.
Lead styphnate [styphnate of lead, lead trinitroresorcinate].
Liquid nitrated polyol and trimethylolethane.
Liquid oxygen explosives.

M
Magnesium ophorite explosives.
Mannitol hexanitrate.
MDNP [methyl 4,4-dinitropentanoate].

MEAN [monoethanolamine nitrate].
Mercuric fulminate.
Mercury oxalate.
Mercury tartrate.
Metriol trinitrate.
Minol-2 [40% TNT, 40% ammonium nitrate, 20% aluminum].
MMAN [monomethylamine nitrate]; methylamine nitrate.
Mononitrotoluene-nitroglycerin mixture.
Monopropellants.

N
NIBTN [nitroisobutametriol trinitrate].
Nitrate explosive mixtures.
Nitrate sensitized with gelled nitroparaffin.
Nitrated carbohydrate explosive.
Nitrated glucoside explosive.
Nitrated polyhydric alcohol explosives.
Nitric acid and a nitro aromatic compound explosive.
Nitric acid and carboxylic fuel explosive.
Nitric acid explosive mixtures.
Nitro aromatic explosive mixtures.
Nitro compounds of furane explosive mixtures.
Nitrocellulose explosive.
Nitroderivative of urea explosive mixture.
Nitrogelatin explosive.
Nitrogen trichloride.
Nitrogen tri-iodide.
Nitroglycerine [NG, RNG, nitro, glyceryl trinitrate, trinitroglycerine].
Nitroglycide.
Nitroglycol [ethylene glycol dinitrate, EGDN].
Nitroguanidine explosives.
Nitronium perchlorate propellant mixtures.
Nitroparaffins Explosive Grade and ammonium nitrate mixtures.
Nitrostarch.
Nitro-substituted carboxylic acids.
Nitrourea.

O
Octogen [HMX].
Octol [75 percent HMX, 25 percent TNT].
Organic amine nitrates.
Organic nitramines.

P
PBX [plastic bonded explosives].
Pellet powder.
Penthrinite composition.
Pentolite.
Perchlorate explosive mixtures.
Peroxide based explosive mixtures.
PETN [nitropentaerythrite, pentaerythrite tetranitrate, pentaerythritol tetranitrate].
Picramic acid and its salts.
Picramide.
Picrate explosives.
Picrate of potassium explosive mixtures.
Picratol.

Picric acid (manufactured as an explosive).
Picryl chloride.
Picryl fluoride.
PLX [95% nitromethane, 5% ethylenediamine].
Polynitro aliphatic compounds.
Polyolpolynitrate-nitrocellulose explosive gels.
Potassium chlorate and lead sulfocyanate explosive.
Potassium nitrate explosive mixtures.
Potassium nitroaminotetrazole.
Pyrotechnic compositions.
PYX [2,6-bis(picrylamino)] 3,5-dinitropyridine.

R
RDX [cyclonite, hexogen, T4, cyclo-1,3,5,-trimethylene-2,4,6,-trinitramine; hexahydro-1,3,5-trinitro-S-triazine].

S
Safety fuse.
Salts of organic amino sulfonic acid explosive mixture.
Salutes (bulk).
Silver acetylide.
Silver azide.
Silver fulminate.
Silver oxalate explosive mixtures.
Silver styphnate.
Silver tartrate explosive mixtures.
Silver tetrazene.
Slurried explosive mixtures of water, inorganic oxidizing salt, gelling agent, fuel, and sensitizer (cap sensitive).
Smokeless powder.
Sodatol.
Sodium amatol.
Sodium azide explosive mixture.
Sodium dinitro-ortho-cresolate.
Sodium nitrate explosive mixtures.
Sodium nitrate-potassium nitrate explosive mixture.
Sodium picramate.
Special fireworks.
Squibs.
Styphnic acid explosives.

T
Tacot [tetranitro-2,3,5,6-dibenzo-1,3a,4,6a tetrazapentalene].
TATB [triaminotrinitrobenzene].
TATP [triacetonetriperoxide].
TEGDN [triethylene glycol dinitrate].
Tetranitrocarbazole.
Tetrazene [tetracene, tetrazine, 1(5-tetrazolyl)-4-guanyl tetrazene hydrate].
Tetrazole explosives.
Tetryl [2,4,6 tetranitro-N-methylaniline].
Tetrytol.
Thickened inorganic oxidizer salt slurried explosive mixture.
TMETN [trimethylolethane trinitrate].
TNEF [trinitroethyl formal].
TNEOC [trinitroethylorthocarbonate].
TNEOF [trinitroethylorthoformate].
TNT [trinitrotoluene, trotyl, trilite, triton].

Torpex.
Tridite.
Trimethylol ethyl methane trinitrate composition.
Trimethylolthane trinitrate-nitrocellulose.
Trimonite.
Trinitroanisole.
Trinitrobenzene.
Trinitrobenzoic acid.
Trinitrocresol.
Trinitro-meta-cresol.
Trinitronaphthalene.
Trinitrophenetol.
Trinitrophloroglucinol.
Trinitroresorcinol.
Tritonal.

U
Urea nitrate.

W
Water-bearing explosives having salts of oxidizing acids and nitrogen bases, sulfates, or sulfamates (cap sensitive).
Water-in-oil emulsion explosive compositions.

X
Xanthamonas hydrophilic colloid explosive mixture.

Approved: October 6, 2011.
B. Todd Jones, Acting Director.

[FR Doc. 2011–26963 Filed 10–18–11; 8:45 am]

www.ingramcontent.com/pod-product-compliance
Lightning Source LLC
Chambersburg PA
CBHW081503170526
45166CB00008B/2530